Springer Theses

Recognizing Outstanding Ph.D. Research

Aims and Scope

The series "Springer Theses" brings together a selection of the very best Ph.D. theses from around the world and across the physical sciences. Nominated and endorsed by two recognized specialists, each published volume has been selected for its scientific excellence and the high impact of its contents for the pertinent field of research. For greater accessibility to non-specialists, the published versions include an extended introduction, as well as a foreword by the student's supervisor explaining the special relevance of the work for the field. As a whole, the series will provide a valuable resource both for newcomers to the research fields described, and for other scientists seeking detailed background information on special questions. Finally, it provides an accredited documentation of the valuable contributions made by today's younger generation of scientists.

Theses may be nominated for publication in this series by heads of department at internationally leading universities or institutes and should fulfill all of the following criteria

- They must be written in good English.
- The topic should fall within the confines of Chemistry, Physics, Earth Sciences, Engineering and related interdisciplinary fields such as Materials, Nanoscience, Chemical Engineering, Complex Systems and Biophysics.
- The work reported in the thesis must represent a significant scientific advance.
- If the thesis includes previously published material, permission to reproduce this must be gained from the respective copyright holder (a maximum 30% of the thesis should be a verbatim reproduction from the author's previous publications).
- They must have been examined and passed during the 12 months prior to nomination.
- Each thesis should include a foreword by the supervisor outlining the significance of its content.
- The theses should have a clearly defined structure including an introduction accessible to new PhD students and scientists not expert in the relevant field.

Indexed by zbMATH.

Francesc Cunillera

Dark Energy

From EFTs to Supergravity

Doctoral Thesis accepted by
University of Nottingham, Nottingham, UK

Springer

Author
Dr. Francesc Cunillera
Centre for Astronomy and Particle Theory
Nottingham Centre for Gravity
University of Nottingham
Nottingham, UK

Supervisors
Prof. Antonio Padilla
Centre for Astronomy and Particle Theory
Nottingham Centre for Gravity
University of Nottingham
Nottingham, UK

Prof. Ed Copeland
Centre for Astronomy and Particle Theory
Nottingham Centre for Gravity
University of Nottingham
Nottingham, UK

ISSN 2190-5053 ISSN 2190-5061 (electronic)
Springer Theses
ISBN 978-3-031-21467-7 ISBN 978-3-031-21468-4 (eBook)
https://doi.org/10.1007/978-3-031-21468-4

This Springer imprint is published by the registered company Springer Nature Switzerland AG
The registered company address is: Gewerbestrasse 11, 6330 Cham, Switzerland

Supervisor's Foreword

There is an enduring debate as to whether the study of string theory, or indeed any theory of quantum gravity, represents science in the traditional sense of making conjectures, deriving predictions and carrying out experiments to test those predictions. The key obstacle is the scale at which quantum gravity effects are likely to manifest themselves, being far in excess of any foreseeable experiment. Nevertheless, such criticisms, at least in the context of string theory, are a little overstated.

String theory is built on two pillars of twentieth-century physics—quantum mechanics and relativity. Indeed, if we consider the two to two scattering of gravitons in General Relativity, the amplitude describing the process violates unitarity—a key ingredient of quantum mechanics—whenever the incident energy exceeds the Planck scale. The only known completion of this amplitude that preserves unitarity, remains weakly coupled and preserves Lorentz invariance to arbitrarily high energies, is the Virasoro-Shapiro (VS) amplitude. This is precisely the amplitude describing the scattering of closed strings in string theory. Thus, the foundation of string theory relies heavily on those two fundamental pillars of quantum mechanics and relativity. If either pillar should fall in experiment, then so will string theory.

In the last decade and a half, the so-called swampland programme has sought to provide another key test of string theory and its underlying principles. What the swampland programme seeks to do is to consider the space of consistent low-energy effective theories and to split them into those that are in the landscape and those that are in the swamp. If a theory is in the swamp, this is because it cannot be derived as the low-energy limit of a consistent compactification of string theory. If it is in the landscape, then the converse is true—we should be able to derive it from string theory. As an example, consider a theory of particles charged under a gauged U(1), minimally coupled to General Relativity. In principle, this theory is a consistent quantum field theory at low energies regardless of the strength of the corresponding electromagnetic and gravitational forces. However, a study of the decay of extremal black holes suggests that only those theories for which gravity is the weakest force can emerge from a consistent theory of quantum gravity like string theory. If the gravitational force is stronger than the electromagnetic force, the theory is expected to be in the swamp. In this sense, we can test quantum gravity simply by comparing the relative strength of gravity compared with the other fundamental forces in nature.

There are many swampland conjectures including the weak gravity conjecture described above, the distance conjecture and the controversial de Sitter conjecture. The latter is the claim that stable de Sitter vacua are in the swamp. The evidence for this is circumstantial, loosely based on the difficulty in deriving consistent solutions from perturbative string compactifications. The underlying reason for this is in the need to break supersymmetry in de Sitter, which can place considerable restrictions on calculational control. However, if true, the de Sitter swampland conjecture has far-reaching implications for both the early and late universe. At early times, the conjecture is in tension with inflation. At late times, it is in tension with dark energy—the observation that the universe is undergoing a period of quasi de Sitter accelerated expansion. To release the tension with late universe physics, it is often assumed that dark energy should not be associated with a de Sitter vacuum, but with a dynamically evolving field dubbed quintessence. This is usually taken to be a scalar in slow roll along a moderately flat (effective) potential.

Cesc's thesis is devoted to understanding whether or not observationally viable quintessence can be derived from string theory. The conclusion of his thesis is remarkable: if we assume the same underlying conditions and assumptions that led to the de Sitter swampland conjecture, then observationally viable quintessence is also in the swamp. Taken at face value, this would be a disaster for string theory. Observations point to a gentle accelerated expansion, whose field theoretic origin is expected to be due to a de Sitter vacuum or quintessence. If string theory is unable to give rise to either, then this points towards a potential conflict with observation. Strings would be ruled out using the scientific method.

A more conservative perspective, and the one that we adopted, was that the problem lies in the swampland programme itself, at least in the context of the de Sitter conjecture. The evidence for the conjecture and its underlying assumptions are not as compelling as for other swampland conjectures, such as the one we described related to weak gravity. Our expectation is that low scale de Sitter vacua and/or quintessence can be derived from string theory, although this may require a better understanding of the physics at strong coupling, far away from the boundary of moduli space.

Cesc's thesis is incredibly important in helping us better understand how the cosmology of the late universe can emerge from fundamental theory. It represents a calculational tour-de-force alongside deep theoretical insights and a panoptic survey of the relevant literature. The projects themselves were driven by Cesc. Although I had a rudimentary knowledge of string theoretic ideas, my previous work had been mainly devoted to field theory. It was Cesc who pushed hard to work on string cosmology, making key breakthroughs and generating exciting collaborations. He has truly been a remarkable student who was operating more like a postdoctoral researcher from the moment he first walked into my office. This thesis and its remarkable results are a fitting testament to his hard work and, of course, his brilliance.

Nottingham, UK
November 2022

Prof. Antonio Padilla

Abstract

The subject of this thesis is cosmological implications of string compactifications understood in a broad sense. In the first half of the thesis, we will begin by reviewing the four-dimensional description of the tree-level perturbative type IIB action. We will then introduce a number of open questions in cosmology and their relevance with regard to the remainder of the thesis. We will first explore some of these questions from the perspective of effective field theories motivated by supergravity. In particular, we provide a description of a naturally light dark energy field in terms of the clockwork mechanism and the Dvali-Kaloper-Sorbo four-form mixing. We study its possible UV completion and show a no-go for its embedding within perturbative type IIA supergravity. We also discuss the coincidence problem for dynamical models of dark energy consistent with a quintessence field slowly rolling down a potential slope, of the type one would expect from the asymptotics of moduli space. As it rolls, a tower of heavy states will generically descend, triggering a phase transition in the low-energy cosmological dynamics after at most a few hundred Hubble times. As a result, dark energy domination cannot continue indefinitely, and there is at least a percentage chance that we find ourselves in the first Hubble epoch.

In the second half of the thesis, we introduce the effects of perturbative and non-perturbative corrections to the tree-level type IIB action. We then focus on obtaining a viable model of quintessence from the type IIB effective field theory. However, we are able to show that such a model must have a non-supersymmetric Minkowski vacuum at leading order. Furthermore, it must necessarily take the form of axion hilltop quintessence. When we consider the effects of quantum fluctuations during the early universe, we see that such models must have extremely fine-tuned initial conditions to describe a slow-rolling scalar field at present times. We conclude that quintessence faces more challenges than a true cosmological constant, to the point that quintessence is very unattractive for model building modulo a ruling out of the cosmological constant by observations. Following this line of reasoning, we consider whether other perturbative corrections can generate de Sitter solutions in an appropriate setting. In particular, we consider the effects of higher curvature corrections in the Gauss–Bonnet term. Remarkably, we are able to show that, for the particular setting of a fluxed runaway potential motivated by heterotic supergravity, the curvature corrections reduce the space of solutions.

Publications Related to This Thesis

The following chapters contain original research presented in the following papers:

- Chapter 4: **A natural theory of dark energy**, Lorenzo Bordin, FC, Antoine Lehébel and Antonio Padilla, *Phys.Rev.D* (2020), arXiv:1912.04905 [hep-th]
- Chapter 5: **A stringy perspective on the coincidence problem**, FC and Antonio Padilla, *JHEP* (2021), arXiv:2105.03426 [hep-th]
- Chapter 6: **Quintessence and the Swampland. The parametrically controlled regime of moduli space.**, Michele Cicoli, FC, Antonio Padilla and Francisco G. Pedro, *Fortsch.Phys.* (2022), arXiv:2112.10779 [hep-th]
- Chapter 8: **Quintessence and the Swampland. The numerically controlled regime of moduli space.**, Michele Cicoli, FC, Antonio Padilla and Francisco G. Pedro, *Fortsch.Phys.* (2022), arXiv:2112.10783 [hep-th]
- Chapter 9: **Quadratic curvature corrections and the absence of de Sitter vacua**, FC, Will Emond, Antoine Lehébel and Antonio Padilla, *JHEP* (2022), arXiv:2112.05771 [hep-th]

All these papers were published under open-access agreements and can be retrieved from the relevant journals.

Acknowledgements

The first and biggest thanks go to Tony. For some reason, you decided that I was worth your time and put up with 3 years of my painstaking mistakes and dumb ideas. You gave me enough freedom to go on a random supergravity journey for my Ph.D. and taught me so much about physics and what it means to be a good physicist during that time. I will always be thankful for all the time you have dedicated to nurture me as a budding physicist, and, hopefully, some of your innate intuition about physics will have rubbed off.

I would also like to give a special thanks to Ed Copeland, Tasos Avgoustidis, Michele Cicoli and Francisco Gil Pedro for many long discussions about strings and physics in general. Ed, you also acted as a supervisor for me, making sure the less obvious admin duties were fulfilled and entertained long discussions about everything cosmological when I would randomly pop into your office. Tasos, our chats and coffees at Choulia's will be one of the things I miss the most from my times at Nottingham. Michele and Francisco, you guys patiently put up with many discussions over Zoom while I was getting up to speed with model building aspects of supergravity. You also made both Tony and I jealous of the beautiful Italian countryside, especially when comparing it to our bleak English weather.

I would also like to thank Arthur Hebecker, Chad Briddon, Joe Conlon, Lorenzo Bordin, Clare Burrage, Will Emond, Cristiano Germani, Oli Gould, Antoine Lehébel, Pete Millington, Adam Moss, Paul Saffin and Marco Scalisi for many useful and insightful discussions over the years.

Contents

Chapter 1
Introduction and Motivations

Let us begin by considering the age-old question: *why string theory?* To answer it, we first consider the two underlying low-energy theories that string theory aims to reproduce at the four-dimensional classical level: the Standard Model (SM) and General Relativity (GR). These two theories have been extremely successful in the many decades since their inception. Almost all of the physical probes[1] currently available to physicists are consistent with a description in terms of GR + SM, from gravitational waves to the low energy particle spectrum of our Universe, at least to scales of order TeV [1]. It is astounding that these two effective low energy theories can explain such a range of physical phenomena to such a degree of accuracy. In fact, one of the current problems within particle physics is the inability to design experiments able to probe regimes where we expect the SM description to break down.[2,3]

Nonetheless, despite the many successes, both the Standard Model and General Relativity falter in some aspects. Regarding the SM, there are issues when one considers it as a theory with some twenty free parameters to be fixed by experiments. Although these parameters can be fixed by experiments, one can ask the following two questions. First, what is the origin of the parameters in the SM Lagrangian? Second, why do some of these parameters take unnatural values, i.e. values that are much smaller than unity? Even if the physical origin and smallness of the parameters in the SM Lagrangian were not a worry, other issues arise when considering quantum

---

[1] As far as we are aware, the only measurement that requires an explicit modification of the SM are the non-zero neutrino masses. The anomalous dipole moment of the muon might also indicate some new physics, but its interpretation is not as clear cut as the three-generations of massive neutrinos, at the moment.

[2] Modulo the discovery of supersymmetry at a collider experiment, given that the supersymmetry breaking scale for the matter sector could lie anywhere between the mass of the top quark and $M_{\rm pl}$.

[3] Between the writing and defence of this thesis, the CDF collaboration [2] claimed a $\sim 7\sigma$ tension between their experimental results and the SM expectation for the mass of the W boson. If confirmed by other experiments, this would be the strongest signal for beyond the SM physics to date.

F. Cunillera, *Dark Energy*, Springer Theses,
https://doi.org/10.1007/978-3-031-21468-4_1

effects [3]. Most notably the Higgs, or any scalar not protected by some symmetry for that matter, is unstable against radiative corrections. This is to say that loop corrections couple the Higgs mass to the heaviest mass available in the theory, one of the incarnations of the so-called hierarchy problem [4]. Therefore, unless new physics comes in to protect the Higgs at some scale above the current collider energies, there would be no reason to expect the Higgs to have its observed mass.

On the other hand, the addition of gravity presents even more problems. Setting aside the usual considerations about singularities of GR, like in the interior of a black hole, the mismatch between theoretical and observational predictions for the value of the vacuum energy density presents one of the biggest open challenges in modern physics [5]. Indeed, in terms of phenomenology we can think of the vacuum energy as a scalar field, which is either a cosmological constant or a very slowly varying scalar dubbed quintessence. Again, as a scalar field which is not protected by a symmetry at the level of GR, the cosmological constant is radiatively unstable and will tend to couple to the heavier masses available in the theory. In this case, the presence of the SM particles alone would naively induce a correction to the cosmological constant of $\mathcal{O}(\text{TeV}^4)$, many orders of magnitude above its observed value at $\mathcal{O}(\text{meV}^4)$ scales. This is the so-called cosmological constant problem [6].

Turning on gravity also introduces technical difficulties, which are not entirely unrelated to the previous issues, in that the quantum field theory of GR+SM becomes non-renormalisable. This is easily seen as soon as one realises that the Ricci term in the Einstein-Hilbert action is parametrising the strength of graviton interactions, with a coupling strength of M_{pl}^{-2}. This is an irrelevant operator, growing weaker at low energies and strong at large energies. It is certainly weak enough that at the level of collider physics one seems to be able to turn gravity off and accurately predict the scattering amplitudes of any interaction. The problem with non-renormalisability becomes clear when considering internal graviton loops. Indeed, in that case the integral over momentum space becomes divergent for arbitrarily high energies. This divergence only becomes worse with higher orders in the loop calculation [5]. This divergence can be mapped to position space as the limit where graviton vertices become coincident [7].

String theory provides a beautiful description of graviton scatterings that gets rid of these divergences. Heuristically, due to the extended nature of strings, the scattering diagrams for gravitons are no longer coincident but spread out over the worldsheet of the string. Incidentally, string theory[4] also has many desirables properties for a would-be unification theory of the SM with gravity. Some of these are [7]:

1. String theory reproduces the low energy description of gravity, i.e. string theory contains a massless spin-2 particle whose interactions reduce to general relativity at the four-dimensional low energy level [8].
2. Supersymmetry of the full theory follows from consistency requirements and it provides a natural way of introducing fermions and bosons into the particle spectrum of the theory, with a very clear physical interpretation in terms of periodic

[4] Henceforth, by string theory we loosely refer to *a complete theory of extended objects as fundamental blocks of nature for which supergravity is a low energy description.*

and anti-periodic boundary conditions on the string worldsheet [9]. It could also aid in solving the SM hierarchy problem and ameliorate the cosmological constant problem, to some extent [3].

3. Further consistency considerations imply that the supersymmetric string theory requires a definite number of spacetime dimensions. Supersymmetric string compactifications allow a low-energy description of the spacetime in terms of 4 large external dimensions, that can host physics similar to those of the SM, and 6 small internal dimensions, that can be *integrated out* at collider energies [10].
4. Supersymmetry can be partially broken so that the low energy dynamics of the string theory are chiral, in agreement with the SM [10].
5. String theory, in its final incarnation as described in the footnote, contains no free parameters and is unique [11].

Of course, the lack of supersymmetry found at the LHC presents some subtleties in engineering a phenomenologically viable low-energy description of string theory. Nonetheless, these issues can, a priori, be fixed by softly breaking supersymmetry in the matter sector at some scale between the energy limit at collider searches and M_{pl}. It is quite striking that the consistency considerations of string theory can lead to such powerful constraints and still yield a theory that is remarkably capable of describing our Universe.

Clearly, string theory represents a powerful framework[5] to unify SM and GR. However, there must be some work left to be done, otherwise theoretical physics would be solved. Most of the open questions in this line of research can be recast as a lack of understanding of the full non-perturbative description of string theory. Indeed, the work in this thesis will concentrate on supergravity, i.e. the low-energy physics of string theory. The supergravity action only contains the massless sector of string theory, that is we will drop the dependence on oscillator modes and keep only the string zero modes. This allows an exploration of the perturbative aspects of string theory, where we see supergravity as an expansion to some order in the genus of the worldsheet, parametrised by the string coupling g_s, and to some order in worldsheet loops, parametrised by the string length l_s, or equivalently by the α'-parameter with $l_s := \sqrt{2\pi\alpha'}$. Thus, supergravity inherits two expansion parameters[6] in g_s and α'. We also note that our knowledge of non-perturbative corrections is quite limited. Nonetheless, when trying to describe the low-energy dynamics of our four-dimensional world, we would expect that in most instances supergravity provides a *close enough description*. Keeping the perturbative expansion under control, so that the supergravity predictions can be trusted, will be a running theme in this thesis.

Within the supergravity approximation, we can begin to ask phenomenological question about our Universe. The central question that we tackle in this work is regarding the microscopic origin of the vacuum energy and its implications. A number of the technical questions about naturalness and hierarchies that concerned the SM,

---

[5] And is possibly the only one.

[6] Note that, on top of the expansion parameters, ten-dimensional supergravities admit a number of fluxes being turned on, which at the low-energy effective level will effectively act as continuous free parameters.

will also apply to the vacuum energy. Indeed, the ten-dimensional supergravity action has a characteristic energy scale of $\mathcal{O}(l_s^{-1})$, with $l_s^{-1} \lesssim M_{\text{pl}}$, whereas the observational dark energy scale lies at $\mathcal{O}(10^{-30} M_{\text{pl}})$. Thus, if supergravity is to explain the scale of dark energy we will require a theory able to generate huge scale hierarchies. Can these large scales hierarchies be generated in the first place? If so, can it be done while avoiding extreme fine-tuning of the supergravity parameters?

These are some important considerations that will play a key role in the latter stages of this thesis. Before moving onto the bulk of our work, we would like to touch upon two more points: EFTs and the Swampland (for a review see [12, 13]). Computations in string theory and, by proxy, supergravity are complex. We have a relatively small toolbox to compute quantities in supergravity in comparison to, say, GR. This means that it is sometimes convenient to explore an idea in a simplified framework within effective field theories that are in a more or less rigorous sense motivated by supergravity. In some sense, by working with supergravity we are doing just so. String theory is too complicated to be treated in full generality, so we concentrate on the massless sector and accept the limitations that this carries. Even then, certain concepts might be far from the reach of our current understanding of supergravity. It is then convenient to study such concepts within effective field theory toy models acknowledging its limitations. When a toy model proves promising, we can then try to find its supergravity embedding.

An immediate question that can be raised is whether any self-consistent EFT inspired by supergravity can find itself embedded in one corner or another of the supergravity parameter space. The reasonable expectation is that this would not be true. For example, one could imagine an EFT with more free parameters than a particular supergravity theory, such that supergravity imposes some non-trivial constraints on the free parameters of the EFT which cannot be satisfied. These constraints are not, in general, accessible from the EFT perspective as they require some knowledge of the UV embedding above the EFT cut-off. The Swampland programme aims to provide conditions for any EFT to be consistently embedded within a UV parent theory.[7] This programme has become an important part of the research done by the community in the past decade. Its more fundamental parts concern M-theory/F-theory compactifications and, in general, have been quite successful in improving our understanding of the ten-dimensional description of supergravity (for example [14–16]). On the other hand, there exists another section of the Swampland programme concerned with deriving EFT constraints from the asymptotics of the ten-dimensional theory. In this regime, the supergravity description becomes quite simplified and strong phenomenological bounds on the resulting EFTs can be imposed [17, 18]. Critically, these results hold only in the asymptotic regime and claiming that they hold anywhere else is a matter of speculation. However, due to the powerful phenomenological constraints, this part of the Swampland has attracted considerable research and statements about the asymptotic behaviour of supergravity have turned into general conjectures anywhere in the parameter space of the theory. In this thesis,

[7] This is usually thought to be string theory/supergravity, although some aspects of the Swampland are claimed to apply more generally.

we will briefly discuss the phenomenological branch of the Swampland programme as some of its results overlap with the research presented here. We will often refer to it as the *Swampland programme* for convenience but we remark that our discussion concerns only this second aspect of the Swampland.

Finally, one could worry whether our current description of our effective low-energetic world, in the SM and GR, is at all consistent with a supergravity embedding. We have already pointed out that string theory has many desirable properties that would make us optimistic that such an embedding exists, we also have constructions that closely resemble the low-energy dynamics of SM+GR, and we have no evidence to the contrary, but one can never disregard the possibility that this could not be the case. However, to conclude that such embedding does not exists with any degree of confidence will require a more complete understanding of supergravity. It is certainly not enough to explore the asymptotics of the ten-dimensional description to claim anything about the bulk of the theory.

1.1 Outline

The contents of this thesis are organised as follows.

In Chap. 2, is a background chapter that serves to explore the fundamentals of Calabi-Yau compactifications from first principles. We begin by motivating the use of Calabi-Yau manifolds and describing their geometry. From there, we build up on their applications to supergravity. We begin by deriving the particle spectrum of type IIB supergravity and, noting that the leftover theory is $\mathcal{N} = 2$ supersymmetric, we motivate the introduction of orientifold planes as a way to break down to the phenomenologically interesting $\mathcal{N} = 1$ supersymmetric theory. In Appendix B, we provide a similar discussion for type IIA supergravity, the dual of type IIB, which will become relevant in subsequent chapters.

Lastly for Chap. 2, we utilise the previous knowledge to describe the effective four-dimensional theory and introduce fluxes, a key ingredient in string phenomenology that allows for a non-flat potential at the four-dimensional level. This chapter represents a full characterisation of perturbative supergravity theories. Calculations that are often omitted in the literature with regards to obtaining the four-dimensional effective field theory are carried out explicitly. Some of the more tedious calculations are moved to Appendix A for ease of discussion. This chapter also includes an introductory discussion regarding the Swampland programme.

With some questions left open regarding corrections to the effective field theory and moduli stabilisation that will be tackled in Chap. 7, we move onto exploring some of the most relevant open research questions in modern physics, and in particular its implications for cosmology. In Chap. 3, we introduce the idea of naturalness, the cosmological constant problem and the coincidence problem, both from a historical and modern perspective. All of these key questions turn out to be intimately related to one another and drive the bulk of the current research in the area. The following two chapters concentrate on studying effective field theories motivated by supergravity that aim at answering some of the questions in Chap. 3. To construct the effective field theories, we make use of the tools described in Chap. 2.

Chapter 4 describes a way of tackling the naturalness dark energy. In this chapter we propose a model of dark energy that is driven by a pseudo-scalar whose light mass emerges naturally from a simple microscopic theory. The ingredients to construct the model are explained in detail and extended in the context of supergravity. Furthemore, we discuss the embedding of this model onto a full string theory and highlight the importance of corrections to the effective theory to accomplish such an embedding.

Chapter 5 reintroduces the coincidence problem and places it in the context of the Swampland programme. We argue that, for string theory compactifications broadly in agreement with the Swampland programme, dark energy is likely to signal the beginning of the end of our Universe as we know it and point to the implications of such theories with respect to the coincidence problem. To characterise this arguments, we build a toy model describing the dynamics of quintessence which is coupled to a tower of heavy states, explicitly demonstrating the breakdown of the cosmological dynamics.

In preparation for the second part of the book, in Chap. 6 we review our findings regarding perturbative supergravity and effective field theories motivate by it. We contrast this conclusions with the claims of the Swampland programme. In doing so, we discuss arguments regarding lack off computational control in supergravity theories, the extend to which we can extract phenomenological information from the Swampland programme and supergravity, and discuss the viability of true parametric control at the asymptotics of moduli space. We conclude that, perturbative and non-perturbative corrections to the type IIB action are key if string theory is to describe our Universe.

In Chap. 7, we retake our analysis of Calabi-Yau compactifications. In this chapter, we endevour to introduce perturbative and non-perturbative effects to the type IIB action. In particular, we concentrate on deriving and motivating the formulae that are commonly used in string phenomenology model building: from leading α' to loop and higher derivative corrections. We derive the form of these corrections and their effect on the four-dimensional potential. Furthermore, we provide a discussion on the validity of the effective description and construct the two largest families of cosmological string models: KKLT (and racetrack) models and LARGE volume scenario models; highlighting the differences among them.

In Chap. 8, we take our knowledge of perturbative and non-perturbative corrections and discuss obstructions to describing dynamical dark energy. The key observation in this chapter is that early—inflation—and late time—dark energy—(quasi-)de Sitter expansions cannot be arbitrarily decoupled, contrary to the usual Wilsonian lore. In fact, by taking into account the constraints that observational and theoretical constraints on inflation pose for dark energy string models, we are able to provide a blueprint of the necessary ingredients that a string phenomenological model must comply with to successfully describe our late Universe. In doing so, we explore different situations in which KKLT, racetrack and LARGE volume scenario models can be constructed finding that only a small subset of them can provide a viable description of our Universe.

Left with few options for model builders, in Chap. 9 we explore the possibility that new corrections could come in to save models that would otherwise clash with observations. In particular, we concentrate on higher curvature corrections in the form of a Gauss-Bonnet term added to the action. This type of correction is well-motivated from heterotic compactifications. We derive the form of the potential for the corrected theory and are able to explicitly calculate a set of necessary and sufficient conditions for the existence of de Sitter solutions. However, for potentials that are best motivated from string theory, this conditions are not satisfied suggesting that higher curvature corrections make it parametrically harder to find de Sitter solutions.

In the last chapter, we conclude by reviewing our findings and provide some general thoughts as to possible directions for future research.

1.2 Further Resources

Although great care has been put into making this book self-contained, a graduate-level knowledge on certain areas is expected. Below we provide these areas together with some references to both quantify the knowledge expected and provide a resource for study in case it is needed.

- General Relativity: Any of the classic books on the subject will provide enough of a basis, for example see: [19] or [20].
- Differential Geometry: Basic knowledge, see: [21] or [22]
- String Theory: Knowledge on the basic computations regarding the bosonic string from [7] and [23], or [24] and [25].
- Cosmology: Knowledge of basic cosmological concept [26].

The following two topics are not strictly necessary, but are likely to be useful to understand the rest of the background material.

- Quantum Field Theory: Flat and curved spacetime calculations, see: [27–29].
- Standard Model: General knowledge, see: [27].

Throughout the book, multiple references are given to the more advanced study topics that might be useful to the reader.

1.3 Units and Conventions

Unless explicitly stated, we take the usual unit conventions where $\hbar = c = M_{\mathrm{pl}} = 1$. We will also set α'=1; this will in turn fix the string scale to $M_s := l_s^{-1} = 1/\sqrt{2\pi}$. For the metric, we take the mostly positive convention with Latin uppercase indices $(M, N...)$ making reference to the ten-dimensional metric, Greek indices $(\mu, \nu, ...)$ referring to the external four-dimensional metric and Latin lowercase indices $(i, j, ...)$

will represent the internal six-dimensional metric. Note that this index convention only refers to metric indices and not to the (co)homology indices which will follow particular conventions set out in the pertinent chapters.

Where ambiguities may arise, we will also "hat" the ten-dimensional quantities, for example $\hat{\phi}$ will be the ten-dimensional dilaton field and ϕ will be its four-dimensional analogue.

References

1. PARTICLE DATA GROUP collaboration (2020) Review of particle physics. PTEP 2020:083C01. https://doi.org/10.1093/ptep/ptaa104
2. CDF Collaboration (2022) High-precision measurement of the W Boson mass with the CDF II detector. Science 376:170. https://doi.org/10.1126/science.abk1781
3. Giudice GF. Naturally speaking: the naturalness criterion and physics at the LHC. https://arxiv.org/abs/0801.2562
4. Gildener E (1976) Gauge symmetry hierarchies. Phys Rev D 14:1667. https://doi.org/10.1103/PhysRevD.14.1667
5. Padilla A. Lectures on the cosmological constant problem. https://arxiv.org/abs/1502.05296
6. Weinberg S (1989) The cosmological constant problem. Rev. Mod. Phys 61:1. https://doi.org/10.1103/RevModPhys.61.1
7. Polchinski J (2007) String theory. Volume 1: An introduction to the bosonic string. Cambridge monographs on mathematical physics. Cambridge University Press, p 12. https://doi.org/10.1017/CBO9780511816079
8. Scherk J, Schwarz JH (1974) Dual models for nonhadrons. Nucl Phys B 81:118. https://doi.org/10.1016/0550-3213(74)90010-8
9. Nilles HP (1995) Phenomenological aspects of supersymmetry. In: Workshop on Gauge theories, applied supersymmetry, and quantum gravity, vol 7, pp 81–105. https://arxiv.org/abs/hep-ph/9511313
10. Ibanez LE, Uranga AM (2012) String theory and particle physics: an introduction to string phenomenology. Cambridge University Press
11. Weigand T (2010) Lectures on F-theory compactifications and model building. Class Quant Grav 27:214004 https://doi.org/10.1088/0264-9381/27/21/214004. https://arxiv.org/abs/1009.3497
12. Palti E (2019) The Swampland: introduction and review. Fortsch Phys 67:1900037. https://doi.org/10.1002/prop.201900037. https://arxiv.org/abs/1903.06239
13. van Beest M, Calderón-Infante J, Mirfendereski D, Valenzuela I. Lectures on the Swampland program in string compactifications. https://arxiv.org/abs/2102.01111
14. Lee SJ, Lerche W, Lockhart G, Weigand T. Holomorphic anomalies, fourfolds and fluxes. https://arxiv.org/abs/2012.00766
15. Kallosh R, Wrase T (2019) dS Supergravity from 10d. Fortsch Phys 67:1800071. https://doi.org/10.1002/prop.201800071. https://arxiv.org/abs/1808.09427
16. Kachru S, Kim M, Mcallister L, Zimet M (2021) de Sitter vacua from ten dimensions. JHEP 12:111. https://doi.org/10.1007/JHEP12(2021)111. https://arxiv.org/abs/1908.04788
17. Obied G, Ooguri H, Spodyneiko L, Vafa C. De Sitter space and the Swampland. https://arxiv.org/abs/1806.08362
18. Bedroya A, Vafa C (2020) Trans-Planckian censorship and the Swampland. JHEP 09:123. https://doi.org/10.1007/JHEP09(2020)123. https://arxiv.org/abs/1909.11063
19. Wald RM (1984) General relativity. Chicago University Press, Chicago, USA. https://doi.org/10.7208/chicago/9780226870373.001.0001
20. Misner CW, Thorne KS, Wheeler JA (1973) Gravitation. W. H. Freeman, San Francisco

21. Schutz BF (1980) Geometrical methods of mathematical physics. Cambridge University Press. https://doi.org/10.1017/CBO9781139171540
22. Nakahara M (2003) Geometry, topology and physics
23. Polchinski J (2007) String theory. Volume 2: Superstring theory and beyond. Cambridge Monographs on Mathematical Physics. Cambridge University Press. https://doi.org/10.1017/CBO9780511618123
24. Green MB, Schwarz JH, Witten E (1988) Superstring theory. Volume 1: Introduction. Cambridge Monographs on Mathematical Physics, vol 7
25. Green MB, Schwarz JH, Witten E (1988) Superstring theory. Volume 2: Loop amplitudes, anomalies and phenomenology, p 7
26. Weinberg S (2008) Cosmology
27. Srednicki M (2007) Quantum field theory. Cambridge University Press, p 1
28. Weinberg S (2005) The quantum theory of fields. Volume 1: Foundations. Cambridge University Press, p 6
29. Wald RM (1995) Quantum field theory in curved space-time and black hole thermodynamics. Chicago Lectures in Physics. University of Chicago Press, Chicago, IL

Part I
String Inspired EFTs

Chapter 2
Perturbative Supergravity in Type IIB Theories

In this chapter we explore the fundamentals of Calabi-Yau compactifications. We begin in Sect. 2.1 by motivating the use and describing the geometry of Calabi-Yau manifolds. In Sect. 2.2, we derive the moduli spectrum and scalar potential for $\mathcal{N} = 2$ type IIB Calabi-Yau compactifications. This leaves us with 2 supersymmetry generators and a non-chiral theory. Phenomenologically, we are interested in a chiral $\mathcal{N} = 1$ supersymmetric theory. To achieve this, in Sect. 2.3, we orientifold our $\mathcal{N} = 2$ action and provide the $\mathcal{N} = 1$ spectrum as well as its effective scalar potential.

Finally, we would like to also discuss how to stabilise the moduli fields of the theory. To do so, in Sect. 2.4, we introduce background fluxes through the three-form flux piece in type IIB. This allows us to stabilise the axio-dilaton and the complex structure sector. The stabilisation of the Kähler sector is a challenging question that we will discuss in detail in Chap. 7, as it requires non-perturbative corrections. We also briefly discuss localised sources and tadpole cancellation conditions. In Sect. 2.5, we discuss the Swampland programme and, in particular, the refined versions of the Swampland distance conjecture and the de Sitter conjecture.

2.1 Special Geometry of Kähler Manifolds

Let $\mathcal{X}$ be a Calabi-Yau (CY) three-fold, i.e. a Kähler manifold in 3 complex dimensions with $SU(3)$ holonomy. By the famous theorem conjectured by Calabi and proved by Yau [1], this description for a CY three-fold is equivalent to that of a Kähler manifold in 3 complex dimensions that is Ricci-flat, since a vanishing first Chern class implies an n-fold Kähler manifold admits a Kähler metric with $SU(n)$ holonomy. A Kähler manifold is a Hermitian manifold with metric $g_{i\bar{j}}$, i.e. the purely holomorphic and anti-holomorphic components of the metric vanish, whose Kähler form

$$J = -ig_{i\bar{j}}dz^i \wedge d\bar{z}^{\bar{j}} \tag{2.1}$$

F. Cunillera, *Dark Energy*, Springer Theses,
https://doi.org/10.1007/978-3-031-21468-4_2

is closed, $dJ = 0$, under the exterior derivative d, where z^i denote complex local coordinates and $\bar{z}^{\bar{\imath}}$ are their complex conjugates.

There are a few properties that make studying CY three-folds attractive. First, the critical superstring lives in a ten-dimensional spacetime. By compactifying a six-dimensional subspace on a CY three-fold, we will be left with an effective four-dimensional description of the low-energy physics of the superstring, matching the low-energy dimensionality of our Universe. Second, a compactification on a CY three-fold gets rid of 3/4 of the underlying supersymmetry of the theory, when compared to a toroidal compactification. The number of preserved supersymmetries is related to the decomposition of the spinor representation $\mathbf{16} \in SO(1, 9)$. The existence of a singlet in the decomposition of the internal spinor representation will lead to the conservation of a supersymmetry in the 4-dimensional spectrum [2]. For a generic compact manifold $\mathcal{M}$ and requiring 4-dimensional Poincaré invariance, the ten-dimensional Lorentz group decomposes into

$$SO(1, 9) \longrightarrow SO(1, 3) \times SO(6), \tag{2.2}$$

and the spinor representation becomes

$$\mathbf{16} \longrightarrow (\mathbf{2}, \mathbf{4}) \oplus (\bar{\mathbf{2}}, \bar{\mathbf{4}}), \tag{2.3}$$

where $\mathbf{2}$ is the Weyl spinor of $SO(1, 3)$ and $\mathbf{4}$ is the corresponding spinor of $SO(6) \cong SU(4)$, with $(\bar{\mathbf{2}}, \bar{\mathbf{4}})$ their respective conjugates. Since $SU(4)$ does not contain a singlet in its representation, this would lead to a non-supersymmetric four-dimensional theory. To preserve some supersymmetry we require $\mathcal{M}$ to have $SU(n)$ holonomy with $n < 4$, that is under parallel transport around a closed loop the spinor singlet comes back to itself with a rotation in $SU(n) \subset SO(6) \cong SU(4)$. For example, in the case of a CY three-fold, which by definition has $SU(3)$ holonomy, the internal spinor representation decomposes as

$$\mathbf{4} \longrightarrow \mathbf{3} \oplus \mathbf{1}, \tag{2.4}$$

such that a singlet is left invariant and an $\mathcal{N} = 1$ ten-dimensional theory compactified on a CY three-fold will preserve $\mathcal{N} = 1$ supersymmetry in four dimensions.[1] Finally, Ricci-flatness, or equivalently $SU(3)$ holonomy, allows to find a one-to-one mapping between metric perturbations and the cohomology of the CY three-folds. We explore this relation in the remainder of this section.

The p-th de Rham cohomology associated to the exterior derivative of the three-fold $H^p(\mathcal{X})$ is defined to be the quotient space

$$H^p(\mathcal{X}) := C^p(\mathcal{X})/Z^p(\mathcal{X}), \tag{2.5}$$

[1] Note that type IIB supergravity has $\mathcal{N} = 2$ in ten dimensions and thus preserves $\mathcal{N} = 2$ supersymmetry after compactification on a CY three-fold. This is by virtue of type IIB having both left and right moving states, which yield a spinor decomposition in terms of $\mathbf{4}_{LR} \oplus \mathbf{4}_{RL}$.

where $C^p(\mathcal{X})$ is the space of closed p-forms, i.e. $dA_p = 0$, and $Z^p(\mathcal{X})$ is the space of exact p-forms, i.e. $A_p = dA_{p-1}$. Then, $H^p(\mathcal{X})$ is composed by the representatives of equivalency classes of p-forms whose elements differ at most by an exact p-form. The dimensionality of the $H^p(\mathcal{X})$ group is given by the Betti number b^p, which is related to the Euler characteristic[2] of $\mathcal{X}$ by

$$\chi = \sum_{p=0}^{6}(-1)^p b^p. \tag{2.6}$$

Since $\mathcal{X}$ is a complex manifold, it will be useful to adopt a holomorphic/anti-holomorphic index notation and to split the p-forms and exterior derivative in holomorphic/anti-holomorphic pieces. Indeed, we will define a (p, q)-form to be a form with p holomorphic and q anti-holomorphic indices

$$A_{p,q} = \frac{1}{p!q!} A_{m1,\ldots,m_p,\bar{n}_1,\ldots,\bar{n}_q} dz^{m_1} \wedge \ldots \wedge dz^{m_p} \wedge d\bar{z}^{\bar{n}_1} \wedge \ldots \wedge d\bar{z}^{\bar{n}_q}. \tag{2.7}$$

We also expand the exterior derivative in Dolbeault operators

$$d = \partial + \bar{\partial}, \quad \partial = dz^a \frac{\partial}{\partial z^a}, \quad \bar{\partial} = d\bar{z}^{\bar{a}} \frac{\partial}{\partial \bar{z}^{\bar{a}}}, \tag{2.8}$$

where $(\partial, \bar{\partial})$ map (p, q)-forms to (p+1,q)-forms and (p, q+1)-forms, respectively. The Dolbeault cohomology[3] group $H^{p,q}(\mathcal{X})$ follows from the definition of the operator[4] ∂. Indeed, $H^{p,q}(\mathcal{X})$ is composed by equivalency classes of closed (p, q)-forms under ∂, whose elements differ at most by an exact (p,q)-form. The dimension of $H^{p,q}(\mathcal{X})$ is the Hodge number $h^{p,q}$ and it follows that

$$b^q = \sum_{p=0}^{q} h^{p,q-p}. \tag{2.9}$$

It is convenient to represent cohomology structure of $\mathcal{X}$ through the so-called the Hodge diamond, since this makes the CY symmetries manifest. The Hodge diamond of any CY three-fold is given by

---

[2] This relation holds even more generally for a CY n-fold $\mathcal{X}_n$ where the Euler characteristic is then given by

$$\chi(\mathcal{X}_n) = \sum_{p=0}^{2n}(-1)^p b^p.$$

[3] In a slight abuse of nomenclature, for the rest of the thesis we will use "cohomology" to refer to either de Rham cohomology or Dolbeault cohomology equally.

[4] Note that the Dolbeault cohomology of $\mathcal{X}$ can be defined with respect to $\bar{\partial}$, since the *two* cohomologies will be mapped one-to-one by the complex conjugation of their elements. This is by virtue of $\mathcal{X}$ being a Kähler manifold.

$$
\begin{array}{ccccccc}
 & & & 1 & & & \\
 & & 0 & & 0 & & \\
 & 0 & & h_{1,1} & & 0 & \\
1 & & h_{1,2} & & h_{1,2} & & 1 \, . \\
 & 0 & & h_{1,1} & & 0 & \\
 & & 0 & & 0 & & \\
 & & & 1 & & &
\end{array}
\tag{2.10}
$$

Note that the Hodge diamond, or the cohomology classes to be more precise, are not a complete characterisation of a Calabi-Yau. Indeed, non-homeomorphic Calabi-Yaus can have the same Hodge diamond. This can be better understood with a simple example. Consider a strip with periodic boundary conditions (a band) and one with anti-periodic boundary conditions (a Möbius strip). Both of them are homotopically equivalent to a circle, and also cohomologically equivalent, however there exists no one-to-one bijective map between the two topologies due to the torsion. Therefore, the two topologies are cohomologically equivalent but not homeomorphic. Although this is far from a proof for generic Calabi-Yaus, we find the intuitive picture useful.

The following symmetries have been exploited to eliminate redundant groups in the Hodge diamond:

- Isomorphism of $H^p(\mathcal{X})$ and $H^{3-p}(\mathcal{X})$ requires $h^{p,0} = h^{3-p,0}$.
- Complex conjugation implies $h^{p,q} = h^{q,p}$.
- Poincaré duality leads to $h^{p,q} = h^{3-p,3-q}$;
- Simply connected CYs will have vanishing first homology group as every one-cycle can be infinitesimally shrunk to a point such that $h^{0,1} = h^{1,0} = 0$.

After applying these symmetries, we can see that the cohomological structure of $\mathcal{X}$ will be determined by a pair of Hodge numbers $h^{1,1}$, $h^{2,1}$ which are related to its Euler characteristic through Eq. (2.9)

$$
\chi = 2\left(h^{1,1} - h^{2,1}\right). \tag{2.11}
$$

One can then introduce a base of harmonic p-forms for each cohomology class, see Table 2.1. We denote the basis of harmonic (1,1)-forms with ω_A, its Poincaré dual being the basis of (2,2)-forms $\tilde{\omega}^A$, with $A = 1, ..., h^{1,1}$. Introducing the basis of $H^3(\mathcal{X})$ to be $\left(\alpha_{\hat{K}}, \beta^{\hat{L}}\right)$, which are related to each other by complex conjugation and $\hat{K}, \hat{L} = 0, ..., h^{1,2}$, the only non-trivial intersection numbers are

$$
\int_{\mathcal{X}} \omega_A \wedge \tilde{\omega}^B = \delta_A^B, \quad \int_{\mathcal{X}} \alpha_{\hat{K}} \wedge \beta^{\hat{L}} = \delta_{\hat{K}}^{\hat{L}}. \tag{2.12}
$$

It is also useful to breakdown the $H^3(\mathcal{X})$ basis in its components as follows $\left(\alpha_{\hat{K}}, \beta^{\hat{L}}\right) = (\alpha_0, \chi_K, \beta^0, \chi^L)$ with $K, L = 1, ..., h^{2,1}$.

Table 2.1 Bases for a Calabi-Yau threefold

Cohomology group	Dimension	Basis
$H^{1,1}$	$h^{1,1}$	ω_A
$H^{2,2}$	$h^{1,1}$	$\tilde{\omega}^A$
H^3	$2+2h^{2,1}$	$\left(\alpha_{\hat{K}}, \beta^{\hat{L}}\right)$
$H^{2,1}$	$h^{2,1}$	χ_K
$H^{3,3}$	1	$\mathrm{dvol}(\mathcal{X})$

At this point, one can expand any p-form of $\mathcal{X}$ as a product of moduli fields and cohomology basis elements. For instance, a 2-form field B_2 can generically be expanded as

$$B_2 = \bar{B}_2(x) + b^A(x)\omega_A, \tag{2.13}$$

where x denotes external coordinates.

Aside from the field content present in the theory, one is free to consider deformations δg_{mn} of the CY metric g_{mn} on $\mathcal{X}$, where the indices (m, n) cover both the holomorphic and anti-holomorphic indices, that is $m = (i, \bar{\imath})$. In particular, we consider deformations that preserve the Ricci flatness of the manifold, i.e. $R_{mn}(g + \delta g) = 0$. In this way, even after deforming the metric, the manifold is guaranteed to be Calabi-Yau [3–5].

What are the set of deformations that preserve the Ricci flatness of $\mathcal{X}$? First of all, notice that if the metric of the manifold g_{mn} is Ricci-flat so will be any other g'_{mn} related to the original metric by diffeomorphisms. This diffeomorphism invariance allows us to gauge fix $\nabla^m \delta g_{mn} = \frac{1}{2}\nabla_n \delta g$, where $\delta g = g^{mn}\delta g_{mn}$, such that the linearised Lichnerowicz equation governing the deformations becomes

$$\Delta \delta g_{mn} := \nabla^p \nabla_p \delta g_{mn} + 2R_{m\ n}^{\ p\ q}\delta g_{pq} = 0, \tag{2.14}$$

and we can study the equations for the pure and mixed indices independently as these decouple.

- Mixed components: The Lichnerowicz equations reads

$$\Delta \delta g_{i\bar{\jmath}} = 0. \tag{2.15}$$

Therefore, $\delta g_{i\bar{\jmath}}$ is a harmonic form of the Laplacian and will admit an expansion in terms of ω_A. These modes are related to non-trivial deformations of the Kähler form J through Eq. (2.1).

- Pure components: In this case, the Lichnerowicz equations reads

$$\Delta \delta g_{ij} = 0, \tag{2.16}$$

implying an expansion in (2,0)-forms. However, since we would like to recover a CY manifold after the deformations, there should exist a one-to-one mapping between the δg_{ij} and some combination of mixed deformations. Indeed, one can generate such a mapping between pure deformations and a new set of mixed component deformations through the unique holomorphic (3,0)-form Ω of $\mathcal{X}$ (cf. Sect. 2.1.2).

All in all, one finds that the moduli space of deformations is spanned by [2, 6]

$$\delta g_{i\bar{j}} = -iv^A(\omega_A)_{i\bar{j}}, \tag{2.17}$$

$$\delta g_{ij} = \frac{i}{||\Omega||^2}\overline{(z^K\chi_K)_{i\bar{i}\bar{j}}}\Omega_{jkl}g^{\bar{i}k}g^{\bar{j}l} := \frac{i}{||\Omega||^2}\bar{z}^K(\bar{\chi}_K)_{i\bar{i}\bar{j}}\Omega^{\bar{i}\bar{j}}_{j}, \tag{2.18}$$

where the last equality in Eq. (2.18) simply introduces shorthand notation and we have defined[5] $||\Omega||^2 := \frac{1}{3!}\Omega_{ijk}\bar{\Omega}^{ijk}$. At this level, the moduli space is locally a direct product of two Kähler manifolds

$$\mathcal{M}^{ks} \times \mathcal{M}^{cs},$$

this will continue being true at the 4D level so the potential generated by either sector will decouple from the other and we can study them independently.[6] Furthermore, this implies that (at least locally) the metrics of both sectors are the metrics of Kähler manifolds independently, i.e. there will exist a pre-potential function for each sector such that the metric can be expressed as a derivative of this pre-potential. In the following two section we explore the structure of $\mathcal{M}^{ks}$ and $\mathcal{M}^{cs}$.

2.1.1 The Kähler Structure of $\mathcal{M}^{ks}$

The first of the two sectors is given by the deformations to the Kähler structure. Intuitively these deformations can be understood as changes in the size of the internal space. For example, in the simple case of a T^2 the Kähler structure deformations are parametrised by the modulus $\rho = iR_1R_2$ where R_i are the two radii of the torus. A change in ρ corresponds to a change in the size of the torus.

Back to the more complicated case of a CY, we have $h^{1,1}$ Kähler moduli. It is convenient to package the moduli coming from the Kalb-Ramond 2-form together with the real scalar moduli of Eq. (2.17) in a new set of imaginary scalars given by

$$t^A := v^A + ib^A. \tag{2.19}$$

[5] Again, in this shorthand notation, this definition really means $||\Omega||^2 := \frac{1}{3!}\Omega_{ijk}\bar{\Omega}_{\bar{i}\bar{j}\bar{k}}g^{\bar{i}i}g^{\bar{j}j}g^{\bar{k}k}$.

[6] Note that the existence of this direct product of two manifolds will be directly related to the no-scale structure of type IIB, as we will discuss in Sect. 2.3.2. Breaking the no-scale structure will mean that the Kähler and complex structure sectors will not decouple anymore.

The metric of the moduli space is defined as

$$G_{AB} := \frac{1}{4\mathcal{V}} \int \omega_A \wedge \star\omega_B = -\frac{1}{4\mathcal{V}} \left(\mathcal{K}_{AB} - \frac{\mathcal{K}_A \mathcal{K}_B}{4\mathcal{V}} \right) = \partial_{t^A} \partial_{\bar{t}^B} K^{\mathrm{ks}}, \tag{2.20}$$

where we find the Kähler potential of the Kähler structure metric K^{ks} to be

$$K^{\mathrm{ks}} = -\ln\left[8\mathcal{V}\right] = -\ln\left[\frac{1}{6}\mathcal{K}_{ABC}(t+\bar{t})^A (t+\bar{t})^B (t+\bar{t})^C\right] \tag{2.21}$$

with $\mathcal{V}$ the volume of $\mathcal{X}$ and we define the triple intersection numbers $\mathcal{K}_{ABC}$ and their *contractions* as

$$\mathcal{K}_{ABC} := \int \omega_A \wedge \omega_B \wedge \omega_C,\ \mathcal{K}_{AB} := \int \omega_A \wedge \omega_B \wedge J = \mathcal{K}_{ABC} v^C,$$

$$\mathcal{K}_A := \int \omega_A \wedge J \wedge J = \mathcal{K}_{ABC} v^B v^C,\ \mathcal{K} := \int J \wedge J \wedge J = \mathcal{K}_{ABC} v^A v^B v^C = 6\mathcal{V}.$$

The metric G_{AB} is fully encoded in a single function K^{ks}. Even more so, the moduli space is a *special Kähler* manifold since the Kähler potential itself is given in terms of a prepopential

$$K^{\mathrm{ks}} = -\ln\left[2(f+\bar{f}) - (t+\bar{t})^A (\partial_{t^A} f + \partial_{\bar{t}^A} \bar{f})\right], \tag{2.22}$$

$$f(t) = -\frac{1}{6}\mathcal{K}_{ABC} t^A t^B t^C, \tag{2.23}$$

To ensure that the metric is positive definite, even after the deformations, we require $\mathcal{K}_{AB}, \mathcal{K}_A, \mathcal{K} > 0\ \forall (A, B)$. These conditions define the Kähler cone spanned by the moduli t^A.

Finally, every *special Kähler* manifold has a complex matrix defined as

$$\mathcal{N}_{\hat{A}\hat{B}} := \bar{\mathcal{F}}_{\hat{A}\hat{B}} + 2i \frac{(\mathrm{Im}\,\mathcal{F})_{\hat{A}\hat{C}}\, t^{\hat{C}} (\mathrm{Im}\,\mathcal{F})_{\hat{B}\hat{D}}\, t^{\hat{D}}}{t^{\hat{C}} (\mathrm{Im}\,\mathcal{F})_{\hat{C}\hat{D}} t^{\hat{D}}}, \tag{2.24}$$

where we use adapted coordinates $t^{\hat{A}} = \{1, t^A\}$, $\mathcal{F} = f(t)$ and $\mathcal{F}_{\hat{A}\hat{B}} = \{0, \partial_{t^A}\partial_{t^B}\mathcal{F}\}$. Plugging in the form of $f(t)$ we obtain

$$\mathrm{Re}\mathcal{N} = \begin{pmatrix} -\frac{1}{3}\mathcal{K}_{ABC} b^A b^B b^C & \frac{1}{2}\mathcal{K}_{ABC} b^A b^B \\ \frac{1}{2}\mathcal{K}_{ABC} b^A b^B & -\mathcal{K}_{ABC} b^C \end{pmatrix}, \tag{2.25}$$

$$\mathrm{Im}\mathcal{N} = -\mathcal{V} \begin{pmatrix} 1 + 4G_{AB} b^A b^B & -4G_{AB} b^B \\ -4G_{AB} b^B & 4G_{AB} \end{pmatrix}, \tag{2.26}$$

$$(\mathrm{Im}\mathcal{N})^{-1} = -\frac{1}{\mathcal{V}} \begin{pmatrix} 1 & b^A \\ b^A & \frac{1}{4}G^{AB} + b^A b^B \end{pmatrix}. \tag{2.27}$$

2.1.2 The Complex Structure of $\mathcal{M}^{cs}$

The second sector is given by the deformations to the complex structure. Back to the analogy with a T^2, these deformations are parametrised by the modulus $\tau = i\frac{R_2}{R_1}$ and control the changes in shape of the torus.

In the case of the CY, there will be $h^{2,1}$ complex structure fields. The metric of the complex structure moduli space is defined as

$$G_{K\bar{L}} := -\frac{\int \chi_K \wedge \bar{\chi}_L}{\int \Omega \wedge \bar{\Omega}}. \tag{2.28}$$

Kodaira's formula gives the basis χ_L in terms of the holomorphic (3,0)-form Ω

$$\chi_K(z,\bar{z}) = \partial_{z^K}\Omega(z) + \Omega(z)\partial_{z^K}\left[-\ln\left(i\int \Omega\wedge\bar{\Omega}\right)\right]. \tag{2.29}$$

The bracket is the Kähler potential for the complex structure metric since

$$G_{K\bar{L}} = \partial_{z^K}\partial_{\bar{z}^L}K^{cs}, \quad K^{cs} = -\ln\left(i\int \Omega\wedge\bar{\Omega}\right) = -\ln i\left(\mathcal{Z}^{\hat{K}}\bar{\mathcal{F}}_{\hat{K}} - \bar{\mathcal{Z}}^{\hat{K}}\mathcal{F}_{\hat{K}}\right). \tag{2.30}$$

Thus, the complex structure moduli space is Kähler as well. In the last relation of Eq. (2.30) we have used that the harmonic (3,0)-form Ω benefits from an expansion in terms of the harmonic bases

$$\Omega(z) = \mathcal{Z}^{\hat{K}}(z)\alpha_{\hat{K}} - \mathcal{F}_{\hat{K}}(z)\beta^{\hat{K}}. \tag{2.31}$$

Since the Kähler potential is only defined up to a complex rescaling of Ω by a holomorphic function

$$\Omega \to \tilde{\Omega} = \Omega e^{-h(z)}, \quad K^{cs} \to \tilde{K}^{cs} = K^{cs} + h + \bar{h}, \tag{2.32}$$

we are allowed to gauge fix $\mathcal{Z}^0 = 1$; this is the so-called Kähler gauge. This leaves $h^{2,1}$ free components for the periods of Ω and one can identify these periods with the z^K moduli fields by defining $z^K = \mathcal{Z}^K/\mathcal{Z}^0$. Furthermore, the manifold $\mathcal{M}^{cs}$ is also a *special Kähler* manifold.

Indeed, one can choose $\mathcal{F} = \frac{1}{2}\mathcal{Z}^{\hat{K}}\mathcal{F}_{\hat{K}}$ and the relations set out in the previous section between *special* manifolds and its *special* matrix still hold (now adapted to the *special* coordinates z^K instead of t^A). Namely[7]

[7] We keep the $\mathcal{Z}^0$ dependence explicit for the sake of generality.

$$K^{\text{cs}} = -\ln i \left|\mathcal{Z}^0\right|^2 \left[2(f-\bar{f}) - (z-\bar{z})^K (\partial_{z^K} f + \partial_{\bar{z}^K}\bar{f})\right], \tag{2.33}$$

$$f(z) = \frac{1}{\left(\mathcal{Z}^0\right)^2}\mathcal{F}, \tag{2.34}$$

and the special matrix (also known in this case as the period matrix) is given by

$$\mathcal{M}_{\hat{K}\hat{L}} := \bar{\mathcal{F}}_{\hat{K}\hat{L}} + 2i\frac{(\text{Im}\,\mathcal{F})_{\hat{K}\hat{M}}\,\mathcal{Z}^{\hat{M}}(\text{Im}\,\mathcal{F})_{\hat{L}\hat{N}}\,\mathcal{Z}^{\hat{N}}}{\mathcal{Z}^{\hat{M}}(\text{Im}\,\mathcal{F})_{\hat{M}\hat{N}}\mathcal{Z}^{\hat{N}}}. \tag{2.35}$$

This matrix will appear in the dimensionally reduced type II supergravity action as the kinetic metric for the complex structure vectors and is most conveniently written down as

$$\int \alpha_{\hat{K}} \wedge \star\alpha_{\hat{L}} = -\left[\text{Im}\,\mathcal{M} + (\text{Re}\,\mathcal{M})(\text{Im}\,\mathcal{M})^{-1}(\text{Re}\,\mathcal{M})\right]_{\hat{K}\hat{L}}, \tag{2.36}$$

$$\int \beta^{\hat{K}} \wedge \star\beta^{\hat{L}} = -\,(\text{Im}\,\mathcal{M})^{-1\,\hat{K}\hat{L}}, \tag{2.37}$$

$$\int \alpha_{\hat{K}} \wedge \star\beta^{\hat{L}} = -\left[(\text{Re}\,\mathcal{M})(\text{Im}\,\mathcal{M})^{-1}\right]_{\hat{K}}^{\hat{L}}. \tag{2.38}$$

With this, we finish our review on the geometry of moduli space. We will now proceed to study the $\mathcal{N}=2$ bosonic sector for the type II supergravity actions.

2.2 Perturbative $\mathcal{N}=2$ Supergravity

The massless bosonic spectrum of type IIB supergravity is composed of two sectors: the NSNS sector comprising of the metric $\hat{g}_{mn}$, the dilaton $\hat{\phi}$ and the Kalb-Ramond two-form $\hat{B}_2$; and the RR sector containing the even p-form gauge fields $\hat{C}_0$, $\hat{C}_2$, $\hat{C}_4$.

Our starting point for this section is the ten dimensional type IIB action in Einstein frame [2]

$$\begin{aligned} S_{IIB} = &-\int \left(\frac{1}{2}\hat{R}\star\mathbf{1} + \frac{1}{4}d\hat{\phi}\wedge\star d\hat{\phi} + \frac{1}{4}e^{-\hat{\phi}}\hat{H}_3\wedge\star\hat{H}_3\right) \\ &-\int \frac{1}{4}\left(e^{-2\hat{\phi}}\hat{F}_1\wedge\star\hat{F}_1 + e^{\hat{\phi}}\hat{F}_3\wedge\star\hat{F}_3 + \frac{1}{2}\hat{F}_5\wedge\star\hat{F}_5 + \hat{C}_4\wedge\hat{H}_3\wedge\hat{F}_3\right), \end{aligned} \tag{2.39}$$

where the field strength are defined as follows

$$\begin{aligned} &\hat{H}_3 = d\hat{B}_2, \quad \hat{F}_1 = d\hat{C}_0, \quad \hat{F}_3 = d\hat{C}_2 - \hat{C}_0\hat{H}_3, \\ &\hat{F}_5 = d\hat{C}_4 - \frac{1}{2}\hat{H}_3\wedge\hat{C}_2 + \frac{1}{2}\hat{B}_2\wedge d\hat{C}_2, \end{aligned} \tag{2.40}$$

and the self-duality of $\hat{F}_5$ must be imposed by hand at the level of the equations of motion.

By making use of the harmonic bases, we can expand the field content of the theory as

$$\begin{aligned}\hat{B}_2 &= B_2(x) + b^A(x)\omega_A, \quad \hat{C}_2 = C_2(x) + c_2^A\omega_A, \\ \hat{C}_4 &= c_4^A(x)\omega_A + \xi^{\hat{K}}(x)\wedge\alpha_{\hat{K}} - \tilde{\xi}_{\hat{K}}(x)\wedge\beta^{\hat{K}} + \rho_A(x)\tilde{\omega}^A. \end{aligned} \tag{2.41}$$

We can further make use of the self-duality of $\hat{F}_5$ to get rid of half of the degrees of freedom of $\hat{C}_4$—we choose to keep the $\xi^{\hat{K}}$ and ρ_A fields. On top of these, we will also obtain a number of moduli fields from the deformations of the metric as discussed in the previous section.

Taking the metric ansatz to be

$$ds^2 = g_{\mu\nu}(x)dx^\mu dx^\nu + g_{i\bar{j}}(x,z)dz^i d\bar{z}^{\bar{j}}, \tag{2.42}$$

where $g_{\mu\nu}$ (with $\mu, \nu = 0, ..., 3$) is the external metric and $g_{i\bar{j}}$ (with $i, \bar{j} = 1, ..., 3$) is the CY metric of $\mathcal{X}$ and plugging the definitions of Eq. (2.40) in Eq. (2.39) one finds the tree-level four-dimensional action[8] [7, 8] (see Appendix A.1 for further details)

$$\begin{aligned} S_{IIB}^{(4)} = \int &-\frac{1}{2}R\star\mathbf{1} + \frac{1}{4}\mathrm{Re}\,\mathcal{M}_{\hat{K}\hat{L}}F^{\hat{K}}\wedge F^{\hat{L}} + \frac{1}{4}\mathrm{Im}\,\mathcal{M}_{\hat{K}\hat{L}}F^{\hat{K}}\wedge\star F^{\hat{L}} \\ &- G_{K\bar{L}}dz^K\wedge\star d\bar{z}^{\bar{L}} - h_{pq}dq^p\wedge\star dq^q, \end{aligned} \tag{2.43}$$

where the real and imaginary parts of $\mathcal{M}_{\hat{K}\hat{L}}$ are found from Eq. (2.35), $F^{\hat{K}} := d\xi^{\hat{K}}$, the complex structure metric $G_{K\bar{L}}$ is given in Eq. (2.30) and we define the four-dimensional dilaton to be

$$e^{\phi} := \frac{1}{\mathcal{V}^{1/2}}e^{\hat{\phi}}, \tag{2.44}$$

Table 2.2 Multiplets for the four-dimensional type IIB supergravity spectrum

Multiplet	Dimension	Field content
Gravity multiplet	1	$(g_{\mu\nu}, \xi^0)$
Vector multiplet	$h^{2,1}$	(ξ^K, z^K)
Hypermultiplet	$h^{1,1}+1$	$(v^A, b^A, c_2^A, \rho_A) + \left(h, \tilde{h}, \phi, C_0\right)$

[8] Where a further Weyl rescaling has been made to reabsorb the internal volume $\mathcal{V}$, so that the action is in Einstein frame.

such that the quaternionic metric h_{pq} becomes[9] [9, 10]

$$\begin{aligned} h_{pq}dq^p \wedge \star dq^q := (d\phi)^2 + G_{AB}dT^A \wedge \star dT^B + \frac{1}{4}e^{2\phi}\mathcal{V}(dC_0)^2 \\ + e^{2\phi}\mathcal{V}G_{AB}\left(dc_2^A - C_0 db^A\right) \wedge \star \left(dc_2^B - C_0 db^B\right) \\ + \frac{1}{16\mathcal{V}}e^{2\phi}G^{AD}\left(d\rho_A - \mathcal{K}_{ABC}c_2^B db^C\right) \wedge \star \left(d\rho_D - \mathcal{K}_{DEF}c_2^E db^F\right) \\ + \frac{1}{4\mathcal{V}}e^{2\phi}\left[dh - \frac{1}{2}\left(\rho_A db^A - b^A d\rho_A\right)\right]^2 \\ + \frac{1}{2}e^{4\phi}\left[d\tilde{h} + C_0 dh + c_2^A d\rho_A + \frac{1}{2}C_0\left(\rho_A db^A - b^A d\rho_A\right)\right. \\ \left. -\frac{1}{4}\mathcal{K}_{ABC}c_2^A c_2^B db^C\right]^2 . \end{aligned} \tag{2.45}$$

In [9, 10], it was shown that the dimensionally reduced type IIB theory in Eq. (2.43) admits a description in terms of a local direct product $\mathcal{M}^Q \times \mathcal{M}^{cs}$, where $\mathcal{M}^Q$ is the quaternionic manifold described by the metric h_{pq} and spanned by the coordinates q^p, with $\mathcal{M}^{ks}$ a submanifold of $\mathcal{M}^Q$, and $\mathcal{M}^{cs}$ the complex structure manifold described in Sect. 2.1.2. The ordered multiplets of Eq. (2.43) are given in Table 2.2. The four-dimensional effective theory described by Eq. (2.43) preserves $\mathcal{N} = 2$ supersymmetry. However, phenomenologically we are interested in $\mathcal{N} = 1$ theories. Half of the supersymmetry can be broken by introducing an orientifold projection in an appropriate manner, as we will discuss in the next section. Furthermore, the scalar potential for the four dimensional scalars is flat. To be able to stabilise the moduli fields we will require the use of background fluxes, which we describe in Sect. 2.4.

2.3 Orientifolding the Supergravity Action

In this section we will briefly discuss the orientifolded truncation of type IIB supergravity. For a more complete description, we point the interested reader to [6, 8, 11] and references therein.

We would like to consider the most generic field content invariant under the action of the $\mathcal{N} = 1$ orientifold operator[10]

$$\mathcal{O} = \Omega_p(-1)^{F_L}\sigma, \tag{2.46}$$

[9] Where we have dualised the (C_2, B_2) gauge fields into the scalars $\left(h, \tilde{h}\right)$, as demonstrated in Appendix A.2.

[10] A second choice of orientifold projection is possible in $\tilde{\mathcal{O}} = \Omega_p\sigma$; this would lead to the introduction of $O5$ and $O9$-planes in the theory.

which is the combined action of the world-sheet parity transformation Ω_p, i.e. the operator that transforms the world-sheet coordinates $(\tau_0, \tau_1) \to (\tau_0, -\tau_1)$, the left-moving fermion number operator $(-1)^{F_L}$ and an isometric and holomorphic involution σ acting solely on $\mathcal{X}$ $(\sigma^2 = \mathbf{1})$. Furthermore, σ leaves the metric and complex structure of $\mathcal{X}$ intact [8]. This implies that the Kähler form will also be invariant under the involution $\sigma^* J = J$, where σ^* is the pullback of σ. Holomorphicity of σ implies that the Dolbeault decomposition for the cohomologies is respected, and in particular $\sigma^* H^{3,0} = H^{3,0}$; together with its idempotency we choose for the holomorphic (3,0)-form Ω

$$\sigma^* \Omega = -\Omega. \tag{2.47}$$

This choice for the orientifold projection is consistent with the introduction of $O3$ and $O7$-planes into the effective action [8].

Another effect of the involution is that the cohomology of $\mathcal{X}$ will split in even and odd cohomologies under its action, by virtue of σ being an involutive symmetry of $\mathcal{X}$, and thus we write

$$H^{p,q} = H_+^{p,q} \oplus H_-^{p,q}, \tag{2.48}$$

such that an element $\omega_\pm \in H_\pm^{p,q}$ transforms like $\sigma^* \omega_\pm = \pm \omega_\pm$. Furthermore, we can see the following [8]

- The involution being isometric means that the metric and orientation of $\mathcal{X}$ are preserved. The Hodge-star operator $\star$ then commutes with the involution and $h_\pm^{1,1} = h_\pm^{2,2}$.
- The holomorphicity of σ implies that there will not be any mixing between holomorphic and antiholomorphic groups $h_\pm^{1,2} = h_\pm^{2,1}$.
- The choice Eq. (2.47) leads to $h_-^{3,0} = h_-^{0,3} = 1$ and $h_+^{3,0} = h_+^{0,3} = 0$.
- The previous two points telll us that $\Omega \wedge \bar{\Omega}$ is invariant under σ^* such that $h_-^{3,3} = h_-^{0,0} = 0$ and $h_+^{3,3} = h_+^{0,0} = 1$.

In Table 2.3 we summarise the notation for the truncated harmonic bases after orientifolding.

Finally, the non-trivial intersection numbers are given by truncation of Eq. (2.12) as

$$\int_{\mathcal{X}} \omega_a \wedge \tilde{\omega}^b = \delta_a^b, \quad \int_{\mathcal{X}} \omega_\alpha \wedge \tilde{\omega}^\beta = \delta_\alpha^\beta, \quad \int_{\mathcal{X}} \alpha_{\hat{k}} \wedge \beta^{\hat{l}} = \delta_{\hat{k}}^{\hat{l}}, \quad \int_{\mathcal{X}} \alpha_\kappa \wedge \beta^\lambda = \delta_\kappa^\lambda. \tag{2.49}$$

The action of the parity operator and the fermion number operator on the ten-dimensional fields is summarised in Table 2.4 [12]. It is clear from it that the invariant orientifolded states must obey

$$\begin{aligned} \sigma^* \hat{\phi} = +\hat{\phi}, \quad \sigma^* \hat{g} = +\hat{g}, \quad \sigma^* \hat{B}_2 = -\hat{B}_2, \\ \sigma^* \hat{C}_0 = +\hat{C}_0, \quad \sigma^* \hat{C}_2 = -\hat{C}_2, \quad \sigma^* \hat{C}_4 = +\hat{C}_4. \end{aligned} \tag{2.50}$$

Table 2.3 Harmonic bases for the orientifolded $\mathcal{X}$

Cohomology group	Dimension	Basis
$H^{1,1}_-$	$h^{1,1}_-$	ω_a
$H^{1,1}_+$	$h^{1,1}_+$	ω_α
$H^{2,2}_-$	$h^{1,1}_-$	$\tilde{\omega}^a$
$H^{2,2}_+$	$h^{1,1}_+$	$\tilde{\omega}^\alpha$
$H^{2,1}_-$	$h^{2,1}_-$	χ_k
$H^{2,1}_+$	$h^{2,1}_+$	χ_κ
H^3_-	$2h^{2,1}_- + 2$	$\left(\alpha_{\hat{k}}, \beta^{\hat{l}}\right)$
H^3_+	$2h^{2,1}_+$	$(\alpha_\kappa, \beta^\lambda)$

Table 2.4 Actions of the parity and fermion number operators on the ten-dimensional spectrum

Field	Under Ω_p	Under $(-1)^{F_L}$	Under $\Omega_p\,(-1)^{F_L}$
$\hat{\phi}$	+	+	+
$\hat{g}$	+	+	+
$\hat{B}_2$	−	+	−
$\hat{C}_0$	−	−	+
$\hat{C}_2$	+	−	−
$\hat{C}_4$	−	−	+

We are now able to expand the ten-dimensional spectrum in the new truncated harmonic bases, much like we did in the $\mathcal{N} = 2$ case. In the following subsections we briefly study the truncation of the $\mathcal{N} = 2$ Kähler manifolds into their $\mathcal{N} = 1$ analogues.

2.3.1 Truncating the Kähler Geometry

After orientifolding, we can expand the Kähler form and the Kalb-Ramond 2-form as[11]

$$\sigma^* J = +J \longrightarrow J = v^\alpha \omega_\alpha \quad \text{and} \quad \sigma^* \hat{B}_2 = -\hat{B}_2 \longrightarrow \hat{B}_2 = b^a \omega_a, \qquad (2.51)$$

where $\alpha = 1, ..., h^{1,1}_+$ and $a = 1, ..., h^{1,1}_-$. The invariance of the Kähler form under orientifolding, together with the condition that the metric must remain semi-positive definite after orientifolding, implies that some intersection numbers have to vanish. In particular, any intersection number (or its contraction) with an odd number of odd

[11] Note that the orientifold invariant states of $\hat{B}_2$ have to be odd under σ^*. However, the external piece $B_2(x)$ is even since σ leaves the external space intact. Therefore, $B_2(x)$ gets projected out by the orientifold projection.

2-cycles will have to vanish since orientifolding will flip the sign of the intersection number, breaking the Kähler cone conditions. This means that

$$\mathcal{K}_{abc} = \mathcal{K}_{a\alpha\beta} = \mathcal{K}_{a\alpha} = \mathcal{K}_{a} = 0. \tag{2.52}$$

Plugging these conditions in the definition of the metric in Eq. (2.20), we find the $\mathcal{N} = 1$ truncated metric for the Kähler sector

$$G_{\alpha\beta} = -\frac{1}{4\mathcal{V}}\left(\mathcal{K}_{\alpha\beta} - \frac{\mathcal{K}_{\alpha}\mathcal{K}_{\beta}}{4\mathcal{V}}\right), \quad G_{ab} = -\frac{1}{4\mathcal{V}}\mathcal{K}_{ab}, \quad G_{\alpha b} = G_{a\beta} = 0, \tag{2.53}$$

where $\mathcal{V} := \frac{1}{6}\mathcal{K}_{\alpha\beta\gamma}v^{\alpha}v^{\beta}v^{\gamma}$ is the volume of the truncated $\mathcal{X}$ and there is no ambiguity in the contractions of the triple intersection numbers since $J \in H_{+}^{1,1}$. It will be convenient for later calculations to also introduce the inverse metrics

$$G^{\alpha\beta} = -4\mathcal{V}\mathcal{K}^{\alpha\beta} + 2v^{\alpha}v^{\beta}, \quad G^{ab} = -4\mathcal{V}\mathcal{K}^{ab}. \tag{2.54}$$

We now concentrate on the complex structure sector. After orientifolding, the unique holomorphic form will be expanded as

$$\sigma^{*}\Omega = -\Omega \longrightarrow \Omega = \mathcal{Z}^{\hat{k}}\alpha_{\hat{k}} - \mathcal{F}_{\hat{k}}\beta^{\hat{k}}, \tag{2.55}$$

where $\hat{k} = 0, ..., h_{-}^{1,2}$. By proxy of Eq. (2.29), the even deformations of the complex structure sector will be projected out, z_k being the only ones left. The Kähler potential for the truncated complex structure sector will become

$$G_{k\bar{l}} = \partial_{z^k}\partial_{\bar{z}^l}K^{\text{cs}}, \quad K^{\text{cs}} = -\ln\left(i\int \Omega\wedge\bar{\Omega}\right) = -\ln\, i\left(\mathcal{Z}^{\hat{k}}\bar{\mathcal{F}}_{\hat{k}} - \bar{\mathcal{Z}}^{\hat{k}}\mathcal{F}_{\hat{k}}\right). \tag{2.56}$$

2.3.2 The Effective $\mathcal{N} = 1$ Action

Let us begin by writing down the orientifold invariant spectrum

$$\begin{aligned} \hat{B}_2 &= b^a(x)\omega_a, \quad \hat{C}_2 = c_2^a\omega_a, \\ \hat{C}_4 &= \xi^k(x)\wedge\alpha_k + \rho_\alpha(x)\tilde{\omega}^\alpha. \end{aligned} \tag{2.57}$$

The truncated effective action is directly inherited from the parent $\mathcal{N} = 2$ action, based on the considerations of the previous subsection,

$$\begin{aligned} S^{(4)}_{IIB,\mathcal{N}=1} = \int & -\frac{1}{2}R\star\mathbf{1} + \frac{1}{4}\text{Re}\,\mathcal{M}_{\kappa\lambda}F^{\kappa}\wedge F^{\lambda} + \frac{1}{4}\text{Im}\,\mathcal{M}_{\kappa\lambda}F^{\kappa}\wedge\star F^{\lambda} \\ & - G_{k\bar{l}}dz^k\wedge\star d\bar{z}^{\bar{l}} - h_{\bar{p}\bar{q}}dq^{\bar{p}}\wedge\star dq^{\bar{q}}, \end{aligned} \tag{2.58}$$

where the quaternionic sector becomes

$$
\begin{aligned}
h_{\tilde{p}\tilde{q}} dq^{\tilde{p}} \wedge \star dq^{\tilde{q}} := (d\phi)^2 + G_{\alpha\beta} dv^\alpha \wedge \star dv^\beta + G_{ab} db^a \wedge \star db^b + \frac{1}{4} e^{2\phi} \mathcal{V} (dC_0)^2 \\
+ e^{2\phi} \mathcal{V} G_{ab} \left(dc_2^a - C_0 db^a \right) \wedge \star \left(dc_2^b - C_0 db^b \right) \\
+ \frac{1}{16\mathcal{V}} e^{2\phi} G^{\alpha\beta} \left(d\rho_\alpha - \mathcal{K}_{\alpha bc} c_2^b db^c \right) \wedge \star \left(d\rho_\beta - \mathcal{K}_{\beta de} c_2^d db^e \right).
\end{aligned} \tag{2.59}
$$

We can see that due to the effect of the orientifold projection, while the z^k coordinates are still the natural candidate that spans $\mathcal{M}^{\text{cs}}$, the coordinates t^A are no longer adequate to describe $\mathcal{M}^{\text{ks}}$, and $\mathcal{M}^{\text{Q}}$ by proxy. Finding the correct choice of coordinates is quite involved and we will only present the result in the following. For $O3/O7$-planes we have [8, 13–15]

- Axio-dilaton: $S = e^{-\phi} - iC_0$.
- Complex structure: $U^k := z^k$, with $k = 1, \ldots, h^{1,2}_-(\mathcal{X})$.
- Odd Kähler moduli: $G^a = c^a - iSb^a$, with $a = 1, \ldots, h^{1,1}_-(\mathcal{X})$.
- Even Kähler moduli:

$$
T_\alpha = \tau_\alpha - \frac{1}{2\left(S + \bar{S}\right)} \mathcal{K}_{\alpha ab} G^a \left(G - \bar{G}\right)^b + i\rho_\alpha, \tag{2.60}
$$

with $\alpha = 1, \ldots, h^{1,1}_+$ and where τ_α are given as an implicit function of v^α by

$$
\tau_\alpha = \frac{1}{2} \int_{\mathcal{X}} \omega_\alpha \wedge J \wedge J = \frac{1}{2} \mathcal{K}_{\alpha\beta\gamma} v^\beta v^\gamma, \tag{2.61}
$$

in this way, τ_α is the volume of a four-cycle $\tilde{\omega}^\alpha \in H^{2,2}_+(\mathcal{X})$, the Poincaré dual to the two-cycle $\omega_\alpha \in H^{1,1}_+(\mathcal{X})$ with volume v^α. We note that in terms of the four-cycles, the volume has become an implicit function of τ_α. In general, we would require to invert the relation $\tau_\alpha = \frac{1}{2}\mathcal{K}_{\alpha\beta\gamma} v^\beta v^\gamma$ but this cannot be done in full generality for a Calabi-Yau with a large number $h^{1,1}$ of Kähler moduli. In [16], some advances have been made in terms of studying Kähler moduli stabilisation in the two-cycle frame allowing for a more general approach to model building with, a priori, any number of Kähler moduli.

We also note that based on the definition of τ_α in Eq. (2.61), we have that the internal volume is a homogeneous function of degree 3/2 on the τ_α, or equivalently on $(T + \bar{T})_\alpha$. This is a powerful result that holds for any CY based on the definition of the internal volume derived from the Kähler geometry. This also implies, due to a nice theorem by Euler, the useful relations

$$
\frac{1}{\mathcal{V}} \frac{\partial \mathcal{V}}{\partial T_\alpha} = \frac{3}{2} \left[\left(T + \bar{T}\right)^{-1} \right]^\alpha, \quad \frac{1}{\mathcal{V}} \frac{\partial^2 \mathcal{V}}{\partial T_\alpha \partial T_\beta} = \frac{3}{4} \left[\left(T + \bar{T}\right)^{-1} \right]^\alpha \left[\left(T + \bar{T}\right)^{-1} \right]^\beta, \tag{2.62}
$$

where $\left[\left(T+\bar{T}\right)^{-1}\right]^{\alpha}$ is taken to be the inverse of $\left(T+\bar{T}\right)_{\alpha}$. The Kähler potential of the truncated theory is then given by

$$K=-\ln\left(S+\bar{S}\right)-\ln\left(-i\int_{\mathcal{X}}\Omega\wedge\bar{\Omega}\right)-2\ln\left(\mathcal{V}\right). \tag{2.63}$$

Here, and in the following, the volume is taken to be in the natural coordinate frame in terms of four-cycles, i.e. $\mathcal{V}=\mathcal{V}(T+\bar{T})$, and the $\mathcal{N}=1$ moduli space is given locally by the direct product

$$\mathcal{M}^{\text{cs}}_{h^{1,2}_{-}}\times\mathcal{M}^{\text{Q}}_{h^{1,1}+1}. \tag{2.64}$$

In these coordinates, the $\mathcal{N}=1$ effective action, in terms of the Kähler form K and superpotential W, takes the well-known form [17, 18]

$$\begin{aligned}\mathcal{S}^{(4)}_{IIB,\mathcal{N}=1}=-\int\frac{1}{2}R\star\mathbf{1}+K_{I\bar{J}}dM^{I}\wedge\star d\bar{M}^{\bar{J}}\\+\frac{1}{4}\text{Re}\,\mathcal{M}_{\kappa\lambda}F^{\kappa}\wedge F^{\lambda}+\frac{1}{4}\text{Im}\,\mathcal{M}_{\kappa\lambda}F^{\kappa}\wedge\star F^{\lambda}+V\star\mathbf{1},\end{aligned} \tag{2.65}$$

where the scalar potential is given by

$$V=e^{K}\left(K^{I\bar{J}}D_{I}WD_{\bar{J}}\bar{W}-3\left|W\right|^{2}\right)+\frac{1}{4}\text{Re}\,\mathcal{M}^{-1\,\kappa\lambda}D_{\kappa}D_{\lambda}, \tag{2.66}$$

where D_{κ} are the would-be D-term contributions to the scalar potential. We have used M^{I} to denote all complex scalars in chiral multiplets in the theory, $K_{I\bar{J}}$ is the Kähler metric obtained from the Kähler potential of Eq. (2.63), i.e. $\partial_{I}\partial_{\bar{J}}K$; and we have introduced the Kähler derivative $D_{I}W=\partial_{I}W+W\partial_{I}K$.

Comparison of the actions in Eqs. (2.58) and (2.65) immediately implies that the scalar potential for the moduli fields is flat, $V=0$, or in other words that the superpotential and D-terms vanish $W=D_{\kappa}=0$. We will have to generate a potential to be able to stabilise the moduli fields. This is no trivial task and, among other things, it will require the introduction of background fluxes in our theory. We will dedicate the next section to the study of flux compactifications. Before doing so, however, it is important to remark a couple of things.

Orientifolding requires the introduction of O-planes. These objects, as well as the presence of D-branes and background fluxes, will backreact onto the Calabi-Yau geometry and the sourced solutions will, in general, not admit a metric ansatz of the form Eq. (2.42). A warp factor must be included to account for the backreaction and the new metric ansatz is [19–21]

$$ds^{2}=e^{2A(z)}g_{\mu\nu}(x)dx^{\mu}dx^{\nu}+e^{-2A(z)}g_{i\bar{j}}(x,z)dz^{i}d\bar{z}^{\bar{j}}, \tag{2.67}$$

In general, the effects of the warping can be dramatic, to the point where the resultant geometry is not necessarily conformally Calabi-Yau anymore. In the remainder of this thesis, however, we will concentrate on compactifications at large volume, relative to the string length, where the warp factor tends to unity and the warped and unwarped metric coincide [20]. In this limit, the formulae derived so far hold. This is the limit of interest, in any case, since we understand supergravity as a low-energy effective theory that is well-defined only in the corner of parameter space with small string coupling and large volume. Furthermore, the perturbative and non-perturbative corrections that will become important in the second half of the thesis will be only under control at large volume. It is important to note that although we use the words "small" and "large" we really mean *finitely* small and *finitely* large, not necessarily the strict limits $g_s \rightarrow 0$ and $\mathcal{V} \rightarrow \infty$ that take us to the boundary of moduli space, more on this later.

Finally, the addition of Op-planes and Dp-branes introduces a term into the ten-dimensional action that is localized in $(p+1)$-dimensions, which we have neglected in Eq. (2.65). This is consistent with considering a small number of $O3/O7$-planes to be coincident with an equal number of $D3/D7$-branes, such that their contributions to Eq. (2.65) cancel locally. At the level of the perturbative action in Eq. (2.65), the main use of localised objects is to satisfy tadpole cancellation conditions, allowing non-trivial background fluxes, and to avoid a number of no-go theorems [22, 23]. We will briefly mention these in some more detail in the next section.

Two other applications that will become relevant in the remainder of the thesis are the introduction of non-perturbative corrections through a stack of N D-branes [22, 24] and the addition of uplifting contributions to the potential through anti $D3$-branes [25]. In the former, a stack of N D-branes will give rise to non-Abelian $U(N)$ gauge theories. The Kähler moduli can then be charged under the action of such non-Abelian groups obtaining a non-flat potential, this will be discussed in more detail in Chap. 7. In the latter, anti D-branes at the tip of a Klebanov-Strassler throat are used to uplift four-dimensional anti-de Sitter (AdS_4) vacua to de Sitter (dS_4) (for example, see [26] for an overview of the uplift mechanism).

Although the ten-dimensional description of such objects is very interesting and of paramount importance, within the scope of this thesis we will only treat these at an effective four-dimensional level. We will therefore not dedicate more time to the ten-dimensional description of localised sources and we point the interested reader to [27–30] for fundamentals and to [31–36] for phenomenological applications.

2.4 Flux Compactifications

As we have seen in the previous section, the scalar potential for the moduli fields is flat. The introduction of background fluxes allows us to generate a non-trivial superpotential at tree-level. The scalar potential obtained from this non-zero superpotential will be able to stabilise the complex structure sector as well as the axio-dilaton. However, we will also show that due to the no-scale structure, the Kähler sector does not receive any contribution from background fluxes and it remains flat. The stabilisation of Kähler moduli will be the motivation to introduce corrections to the tree-level action in the second half of the thesis. Finally, let us note that the introduction of fluxes provides a way to spontaneously break the leftover $\mathcal{N} = 1$ supersymmetry.

Take a p-form gauge field A_p with field strength $F_{p+1} = dA_p$. We will say, by analogy with the case of electromagnetism in 4-dimensions, that the gauge field A_p gives rise to a *magnetic flux* of the form

$$\int_{\Sigma_{p+1}} F_{p+1} = m \in \mathbb{Z}, \tag{2.68}$$

where $\Sigma_{p+1} \subset \mathcal{X}$ is the (p+1)-cycle wrapping the (p+1)-form field strength F. The same gauge field will also give rise to an *electric flux* of the form

$$\int_{\Sigma_{D-p-1}} \star F_{p+1} = n \in \mathbb{Z}, \tag{2.69}$$

where $D = \dim(\mathcal{X})$ and $\Sigma_{D-p-1} \in \mathcal{X}$ is the (D-p-1)-cycle wrapping the electric dual of F. We note that these fluxes are quantised in units of $\frac{1}{2\pi\alpha'}$ as a consequence of a generalised Dirac quantisation condition [20], much like the electric and magnetic fluxes are quantised in the case of electromagnetism. However, within the context of the low energy supergravity description, these fluxes will be treated as continuous parameters in the effective theory.

In type IIB supergravity, one can turn on background fluxes for the 3-form field strengths in the NSNS and RR sector. For the background fluxes to be consistent with our orientifold projection we see that the flux pieces must be odd under the action of the involution σ. Thus, we expand the flux pieces as

$$\tilde{H}_3 = m^{\hat{k}}_{(H)}\alpha_{\hat{k}} + n^{(H)}_{\hat{k}}\beta^{\hat{k}}, \quad \tilde{F}_3 = m^{\hat{k}}_{(F)}\alpha_{\hat{k}} + n^{(F)}_{\hat{k}}\beta^{\hat{k}}, \quad \hat{k} = 0, \ldots, h^{1,2}_{-}, \tag{2.70}$$

and the fluxed effective action can be obtained by substituting [8]

$$\hat{H}_3 \to \hat{H}_3^{(F)} = \hat{H}_3 + \tilde{H}_3, \quad \hat{F}_3 \to \hat{F}_3^{(F)} = \hat{F}_3 + \tilde{F}_3, \tag{2.71}$$

in the ten-dimensional action since the $\hat{F}_5$ terms that would involve the 3-form fluxes are projected out due to the orientifold constraints. It is convenient to repackage the two 3-form field strength in the imaginary self-dual three-form field strength [20]

$$\hat{G}_3 = \hat{F}_3^{(F)} - iS\hat{H}_3^{(F)}, \quad \star\hat{G}_3 = i\hat{G}_3, \tag{2.72}$$

and we define the shorthand notation

$$\tilde{G}_3 = (m_{(F)} - iSm_{(H)})^{\hat{k}}\alpha_{\hat{k}} + (n_{(F)} - iSn^{(H)})_{\hat{k}}\beta^{\hat{k}} := m^{\hat{k}}\alpha_{\hat{k}} + n_{\hat{k}}\beta^{\hat{k}}. \tag{2.73}$$

The fluxed scalar potential is given by [6, 8, 20, 37–39]

$$V = e^K \left(\int \Omega \wedge \bar{\tilde{G}}_3 \int \bar{\Omega} \wedge \tilde{G}_3 + G^{kl} \int \chi_k \wedge \tilde{G}_3 \int \bar{\chi}_l \wedge \bar{\tilde{G}}_3 \right). \tag{2.74}$$

From this expression, it is clear to see that setting the background fluxes to zero will make the scalar potential vanish. To make a connection with Eq. (2.65), it is also shown, for orientifolds without an odd-Kähler sector, the GVW superpotential

$$W_{GVW} = \int \Omega \wedge \tilde{G}_3, \tag{2.75}$$

generates the same scalar potential when making use of Eq. (2.66). For the case of orientifold with an odd Kähler sector this same results holds as long as the no-scale structure is unbroken. Generating a potential for the Kähler sector directly relates to breaking this no-scale structure, since (at least) the volume mode will couple to the other sectors as well as the Kähler moduli. Given that our objective is to discuss four-dimensional phenomenology, we will have to eventually break no-scale. In the remainder of this thesis, we will concentrate in orientifolds that project out the odd Kähler sector, i.e. $h^{1,1} = h^{1,1}_+$, and the results discussed so far will apply straightforwardly. Now that we have generated a scalar potential, we would like to see what sectors can be stabilised from it. As already anticipated, only the axio-dilaton and complex structure moduli will receive a non-zero potential. Let us see this now.

The idea is to show that the scalar potential obtained from GVW superpotential is independent of the Kähler coordinates T_α Eq. (2.60), up to an overall factor. We begin by noting that

$$W_{GVW} = W_{GVW}(S, U), \tag{2.76}$$

since $\Omega = \Omega(U)$ and $\tilde{G}_3 = \tilde{G}_3(S)$. The F-terms for T_α are

$$D_T W_{GVW} = W_{GVW} K_T = -2 W_{GVW} \frac{\mathcal{V}_T}{\mathcal{V}} = -3 W_{GVW} \left[(T + \bar{T})^{-1} \right]^\alpha, \tag{2.77}$$

where $X_T := \frac{\partial X}{\partial T_\alpha}$, we have also made use of the form of the Kähler potential given in Eq. (2.63). The Kähler metric will be block diagonal in the (S, U^k, T_α) sectors such that the scalar potential takes the form

$$\begin{aligned} V &= e^K \left[K^{S\bar{S}} D_S W D_{\bar{S}} \bar{W} + K^{U\bar{U}} D_U W D_{\bar{U}} \bar{W} + K^{T\bar{T}} D_T W D_{\bar{T}} \bar{W} - 3 |W|^2 \right] \\ &= e^K \left[K^{S\bar{S}} D_S W D_{\bar{S}} \bar{W} + K^{U\bar{U}} D_U W D_{\bar{U}} \bar{W} + \left(K^{T\bar{T}} \partial_T K \partial_{\bar{T}} K - 3 \right) |W|^2 \right], \end{aligned} \tag{2.78}$$

where we remark that the first two terms in the brackets are independent of T_α.

The Kähler metric for the T_α sector and its inverse are given by

$$K_{T\bar{T}} = \frac{2}{\mathcal{V}}\left(\frac{\mathcal{V}_T\mathcal{V}_{\bar{T}}}{\mathcal{V}} - \mathcal{V}_{T\bar{T}}\right) = 3\left[(T+\bar{T})^{-1}\right]^{\alpha}\left[(T+\bar{T})^{-1}\right]^{\beta}, \tag{2.79}$$

where we have made use of the relations in Eq. (2.62). Contracting the inverse metric we find

$$K^{T\bar{T}}K_T K_{\bar{T}} = 3, \tag{2.80}$$

so that the parentheses in Eq. (2.78) vanish identically and the scalar potential has the no-scale structure

$$V = e^K\left(K^{S\bar{S}}D_S W D_{\bar{S}}\bar{W} + K^{U\bar{U}}D_U W D_{\bar{U}}\bar{W}\right), \tag{2.81}$$

with the only T_α dependence in the overall exponential factor. From the previous expression, it is clear that the Kähler sector remains flat at tree-level.

Breaking the no-scale structure can be done in one of two ways, either the superpotential gains a dependence on the Kähler moduli $W(S,U) \to W(S,U,T)$ or by breaking the block diagonal structure of the Kähler metric. The first option will be related to the introduction of non-perturbative corrections to the superpotential. The second one can be associated to perturbative corrections in the large volume/small coupling expansion. We will dedicate Chap. 7 to the study of these effects.

By looking at Eq. (2.81) we see that, since any supersymmetric vacua must have vanishing F-terms $D_U W = D_S W = 0$, only supersymmetric Minkowski vacua are allowed. Non-supersymmetric solutions with non vanishing F-terms are also allowed by appropriate choices of fluxes. We can see this by turning once again to the F-terms.

We already found the F-term for the Kähler sector in Eq. (2.77). Preserving supersymmetry will require $W_{\text{GVW}} = 0$, in other words the (0,3)-component of $\tilde{G}_3$ must vanish. The remaining F-terms read

$$D_S W = \frac{1}{S-\bar{S}}\int_{\mathcal{X}} \Omega \wedge \bar{\tilde{G}}_3 = 0, \quad D_U W = \int_{\mathcal{X}} \chi_k \wedge \tilde{G}_3 = 0. \tag{2.82}$$

These F-terms imply that the (3,0)-component and the (1,2)-component need to vanish. Therefore, we have the following: if $\tilde{G}_3 \in H^{2,1}_-$ all conditions are satisfied and supersymmetry is preserved, if $\tilde{G}_3 \in H^{0,3}_-$ we find some broken supersymmetry in $D_T W \neq 0$ with $V = 0$, and if $\tilde{G}_3 \in H^{3,0}_- \oplus H^{1,2}_-$ none of the above F-terms would vanish and we only find unstable vacuum configurations.

Since the axio-dilaton and the complex structure sector are fixed at tree-level, one is safe to consider them fixed by their vacuum expectations values. We will take the tree-level superpotential to be

$$W_{\text{tree}} := W_0 = \left\langle \int_{\mathcal{X}} \Omega \wedge \tilde{G}_3 \right\rangle, \tag{2.83}$$

for the remainder of the thesis, effectively integrating out the axio-dilaton and the complex structure moduli. Any quantum corrections to their vevs in the large volume limit will be subleading to the tree-level contributions and we will neglect them safely. This also implies that the tree-level Kähler potential will take the form

$$K_{\text{tree}} := -2\ln(\mathcal{V}) + K_0 = -2\ln(\mathcal{V}) - \ln\left(\frac{2}{g_s}\right) - \ln\left\langle -i\int_{\mathcal{X}} \Omega \wedge \bar{\Omega}\right\rangle. \quad (2.84)$$

Finally, before moving on, we would like to discuss the tadpole cancellation conditions in the presence of fluxes and localised sources. The choice of non-trivial fluxes induces a new term in the Bianchi identity for the $\hat{F}_5$ field strength. In particular,

$$d\hat{F}_5 = \tilde{H}_3 \wedge \tilde{F}_3, \quad (2.85)$$

which leads to the no-go theorem of [40]. Since the fluxes are the only contribution to the $\hat{C}_4$ tadpole, one would have to consider compactifications with trivial fluxes only. However, the discovery of D-branes [41] led to the realisation that charge densities from $D3/D7$ branes, and also from orientifold planes, contribute to the tadpole [20]. To satisfy the Bianchi identity and guarantee the absence of anomalies in the effective four-dimensional theory, it is then required that the following tadpole cancellation condition holds true

$$N_{D3} - N_{\overline{D3}} + \int_{\mathcal{X}} \tilde{H}_3 \wedge \tilde{F}_3 = \frac{\chi(\mathcal{X}_4)}{24}, \quad (2.86)$$

where $\left(N_{D3}, N_{\overline{D3}}\right)$ are the charge contributions from $D3$-branes and $\overline{D3}$-branes, respectively, and $\chi(\mathcal{X}_4)$ is the contribution from $D7$-branes and orientifold planes, which gains a geometrical interpretation in terms of the Euler characteristic of the corresponding four-fold $\mathcal{X}_4$ [42].

2.5 The Swampland Programme

In the last section we achieved the stabilisation of the non-Kähler moduli through background fluxes. We highlighted that stabilising the Kähler sector will generically require the introduction of perturbative and non-perturbative correction to the tree-level Kähler potential and superpotential discussed previously. These corrections are quintessential to supergravity model building, as we will explore in Chap. 7. It turns out that obtaining a systematic understanding of the stabilisation of the Kähler moduli is a highly non-trivial task. Furthermore, since we lack a complete picture of non-perturbative supergravity, it is complicated to define frameworks where computations can be carried out under complete control. Defining *complete control* is itself a hotly

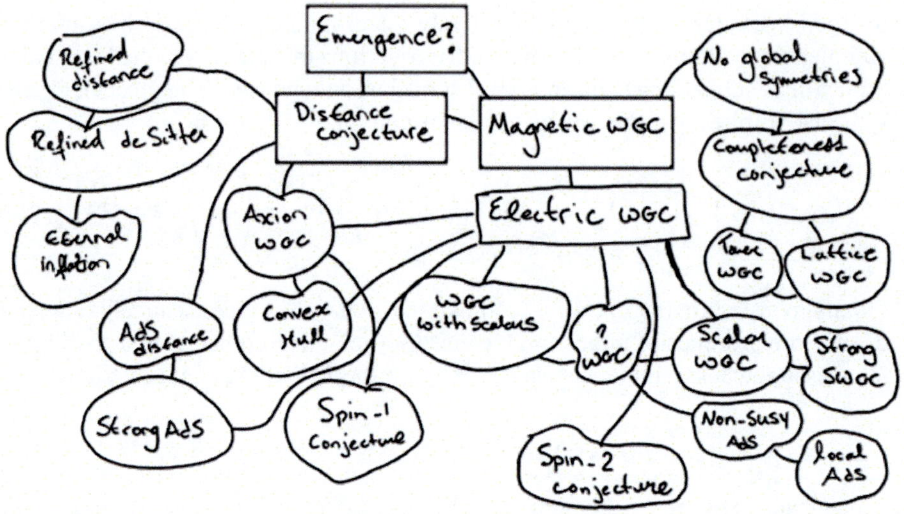

Fig. 2.1 A pictorial representation of the web of conjectures in the Swampland programme [48]

debated topic in the community. Another reason behind the difficulty of the task at hand is that supergravity inherits two perturbation parameter from string theory: α' and the string coupling g_s. In writing down quantum corrections to the tree-level action, we will inevitably encounter terms of $\mathcal{O}(\alpha^p \cdot g_s^q)$, for some $p, q > 0$. This can make it so that, at times, it can become unclear which terms enter at a given order in perturbation theory. For example, comparing terms $\mathcal{O}(\alpha'^3 g_s^2)$ and $\mathcal{O}(\alpha'^2 g_s^3)$ can be hard. A very important part of ongoing supergravity research is to systematically analyse these corrections at all orders and ensure that Kähler moduli stabilisation is not spoiled in the bulk of moduli space (see for example [26, 43–46]).

The lack of a completely satisfactory answer to the challenges presented above led to the creation of the so-called *Swampland programme* (see [47] for an in-depth review). Its core goal is to differentiate between low energy effective field theories that have a consistent completion within a prospective quantum gravity UV theory and those that do not. In doing so, the Swampland programme has formed a *web of conjectures* (Fig. 2.1), that attempt to delineate the space between the consistent and inconsistent theories. In the next section we will briefly describe two of these conjectures, the particular ones that concern the contents of this thesis. One should remark that, although the ultimate goal is for the programme to be independent of a prospective UV completion, most of the conjectures will draw from string theoretical intuitions.

2.5.1 The Swampland Distance Conjecture

In [49, 50], the authors proposed that a number of conjectures. Among them was what would eventually become to be known as the Swampland distance conjecture (SDC). A refined version of the SDC (rDC) claims the following [47, 51, 52],

Conjecture 1 *Consider a theory coupled to gravity with some moduli space parametrised by the expectation values of some fields with no potential. Let the geodesic distance between any two points $P \in \mathcal{M}$ and $Q \in \mathcal{M}$ in moduli space be denoted by $d(P, Q)$. Then, there exists a tower of states in theory with mass scale M and scaling*

$$M(Q) < M(P)e^{-\alpha \frac{d(P,Q)}{M_{pl}}}, \tag{2.87}$$

if $d(P < Q) \gtrsim M_{pl}$ and where $\alpha \sim \mathcal{O}(1)$ is a positive model dependent coefficient.

A second claim of the rDC is that this statement holds true even for fields with a potential by exchanging the moduli space with the field space of the effective field theory.

The motivation for the existence of a light tower of states can be understood from the mass spectrum in D dimensions of a closed string when compactifying one direction on a circle of radius R. Indeed, the $(D-1)$-dimensional mass for the for the string is [53]

$$M^2 \sim n^2 \left(\frac{1}{R^2}\right)^{\frac{d-1}{d-2}} + w^2 \left(\frac{R^2}{\alpha'^2}\right)^{\frac{d-1}{d-2}}, \tag{2.88}$$

where $(n, w) \in \mathbb{N}$ are the Kaluza-Klein number and winding number, respectively. Indeed, in the infinite distance limit $R \to \infty$, a infinite tower of Kaluza-Klein modes labelled by n will become massless. The existence of the winding modes ensures that in the T-dual frame, where $R \to \tilde{R} = 1/R$, there also exists another tower becoming massless, in this case related to the winding modes with labels w.

More detailed analysis of the conjecture has been carried out in a number of string scenarios [51, 52, 54–59]. Generically, the analyses focus on finding solutions and then identifying the objects in the theory that introduce an infinite tower of states which becomes light at the boundary of moduli space. For the volume modulus, it is quite straightforward to show that the tower of states can be associated with the KK modes like we did in the simple example above. In more generality, the tower of states can be harder to identify and related to, for example, D-branes becoming tensionless.

2.5.2 The Refined de Sitter Conjecture

A second conjecture is connected to the existence of (meta)stable dS backgrounds in an effective field theory obtained from string theory compactifications. In [60], the authors introduced the following claim that a potential of any theory coupled to gravity would have to satisfy

$$|\nabla V| \geq \frac{c}{M_{\rm pl}} V. \tag{2.89}$$

It was pointed out in [61], that this claim was in tension with the Higgs potential. The previous statement was refined to its present form in [62, 63]. This refined de Sitter conjecture (rdSC) claims the following

Conjecture 2 *Given an effective field theory consistent with a theory of quantum gravity, its potential must satisfy either*

$$|\nabla V| \geq \frac{c}{M_{pl}} V, \quad or \quad min\left(\nabla_i \nabla_j V\right) \geq -\frac{c'}{M_{pl}^2} V, \tag{2.90}$$

where $(c,\ c')$ *are positive constants of* $\mathcal{O}(1)$ *and min* $\left(\nabla_i \nabla_j V\right)$ *is the minimum of the eigenvalue of the corresponding Hessian.*

This conjecture forbids the existence of stable de Sitter vacua in string theory. Importantly, metastable vacua are allowed as long as a tachyonic instability exists such that the lifetime of the vacua is bounded by its Hubble time $H^{-1} \sim M_{\rm pl}/\sqrt{V}$.

Whereas the refined distance conjecture has been well studied, the UV origin of the refined de Sitter conjecture is more opaque. In part, this is related to the fact that the de Sitter conjecture is entirely supported by cherry-picked examples and difficulty of obtaining de Sitter vacua in string theory. To some extent, it has been argued that the rdSC is consistent with the rSDC [63], and thus the rdSC could be understood from an infinite distance limit perspective in the boundary of moduli space.

If the statement of the de Sitter conjecture was the inability of obtaining deSitter vacua in the boundary of moduli space, this would not be a surprise. Indeed, it has been known for quite some time [64] that non-supersymmetric degrees of freedom (like those of de Sitter vacua) are notably hard to access from the point of view of weak coupling (perturbative) supergravity. In the infinite distance limit all we have left are perturbative degrees of freedom. Therefore, it would be naïve to expect that de Sitter vacua are hiding in that corner of parameter space. However, the dS conjecture forbids dS vacua existing also in the bulk of moduli space, where perturbative and non-perturbative corrections become important.

Indeed, let us consider a heuristic scenario to exemplify this point—we will make this point more formal in Chap. 6 and when discussing corrections to the tree-level action in Chap. 7. In [64], the authors describe the Dine-Seiberg Minkowski vacua, the decompactification limit, for the dilaton modulus. A similar argument for the canonical volume modulus, φ, can also be made. Indeed, let us imagine that we have some de Sitter vacua generated through corrections to the tree-level action as depicted in Fig. 2.2.

At large field values, the potential decreases exponentially tending to the Dine-Seiberg type vacuum, in the language of the canonical field φ. This behaviour at

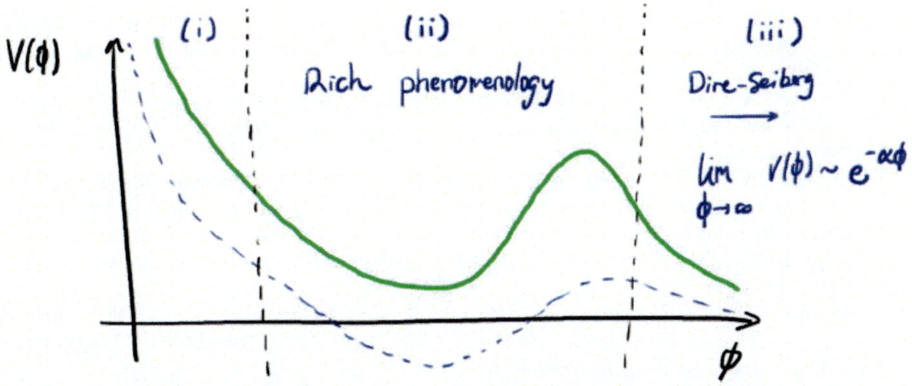

Fig. 2.2 A heuristic example of a typical profile for the scalar potential of a canonical saxion modulus ϕ. The potential consists of three regions: (i) the deep bulk of moduli space where the perturbative expansion breaks down, (ii) the asymptotic bulk where numeric control can be retained and (iii) the boundary of moduli space where the perturbative expansion can be made exact

large field values is quite generic since, in this limit, we expect the terms in the action Eq. (2.58) to dominate over any corrections, and the non-canonical saxions (s) enter the theory as $d\phi \sim s^{-1}ds$, based on the form of the block diagonal Kähler metric obtained from Eq. (2.63).

We can observe two things. Moving the minimum in the bulk towards asymptotic larger values of the field would imply tuning the corrections to the point where the leading perturbative expansion is comparable with the tree-level terms, at which point we lose computational control. In Chap. 6 we revisit the discussion of the Swampland programme from the point of view of phenomenology and point out that there exists a disconnect between the Swampland programme and phenomenology. Secondly, an interesting phenomenological region exists in the asymptotic bulk of moduli space, where the effective theory can remain under numeric control. That region is where all the known de Sitter proposals exists and where the corrections to the tree-level action remain under control. This is where we will concentrate our efforts in the second half of the thesis. Finally, a third region exists, the deep bulk of moduli space, where the effective theory becomes strongly coupled and the perturbative expansion breaks down. This regions maps to small volumes (in units of string length) and/or a string coupling that becomes larger than unity.

References

1. Yau S-T (1977) Calabi's Conjecture and some new results in algebraic geometry. Proc Nat Acad Sci 74:1798. https://doi.org/10.1073/pnas.74.5.1798
2. Polchinski J (2007) String theory. Vol. 2: Superstring theory and beyond. Cambridge monographs on mathematical physics, Cambridge University Press. https://doi.org/10.1017/CBO9780511618123

3. Grimm TW, Louis J (2004) The Effective action of N = 1 Calabi-Yau orientifolds. Nucl Phys B 699:387. https://doi.org/10.1016/j.nuclphysb.2004.08.005. https://arxiv.org/abs/hep-th/0403067
4. Grimm TW, Louis J (2005) The effective action of type IIA Calabi-Yau orientifolds. Nucl Phys B 718:153. https://doi.org/10.1016/j.nuclphysb.2005.04.007. https://arxiv.org/abs/hep-th/0412277
5. Grimm TW (2005) The effective action of type II Calabi-Yau orientifolds. Fortsch Phys 53:1179. https://doi.org/10.1002/prop.200510253. https://arxiv.org/abs/hep-th/0507153
6. Grimm TW, Louis J (2004) The effective action of N = 1 Calabi-Yau orientifolds. Nucl Phys B 699:387. https://doi.org/10.1016/j.nuclphysb.2004.08.005. https://arxiv.org/abs/hep-th/0403067
7. Bohm R, Gunther H, Herrmann C, Louis J (2000) Compactification of type IIB string theory on Calabi-Yau threefolds. Nucl Phys B 569:229. https://doi.org/10.1016/S0550-3213(99)00796-8. https://arxiv.org/abs/hep-th/9908007
8. Grimm TW, The effective action of type II Calabi-Yau orientifolds. Fortsch Phys 53:1179. https://doi.org/10.1002/prop.200510253. https://arxiv.org/abs/hep-th/0507153
9. Ferrara S, Sabharwal S (1989) Dimensional reduction of type II superstrings. Class Quant Grav 6:L77. https://doi.org/10.1088/0264-9381/6/4/002
10. Ferrara S, Sabharwal S (1990) Quaternionic manifolds for type II superstring vacua of Calabi-Yau spaces. Nucl Phys B 332:317. https://doi.org/10.1016/0550-3213(90)90097-W
11. Grimm TW, Louis J (2005) The effective action of type IIA Calabi-Yau orientifolds. Nucl Phys B 718:153. https://doi.org/10.1016/j.nuclphysb.2005.04.007. https://arxiv.org/abs/hep-th/0412277
12. Dabholkar A (1997) Lectures on orientifolds and duality. In: ICTP Summer school in high-energy physics and cosmology, vol 6, pp 128–191. https://doi.org/10.1016/j.nuclphysb.2005.04.007. https://arxiv.org/abs/hep-th/9804208
13. Haack M, Louis J (2001) M theory compactified on Calabi-Yau fourfolds with background flux. Phys Lett B 507:296. https://doi.org/10.1016/S0370-2693(01)00464-6. https://arxiv.org/abs/hep-th/0103068
14. Becker K, Becker M (2001) Supersymmetry breaking, M theory and fluxes. JHEP 07:038. https://doi.org/10.1088/1126-6708/2001/07/038. https://arxiv.org/abs/hep-th/0107044
15. Cicoli M (2010) String loop moduli stabilisation and cosmology in IIB flux compactifications. Fortsch Phys 58:115. https://doi.org/10.1002/prop.200900096. https://arxiv.org/abs/0907.0665
16. AbdusSalam S, Abel S, Cicoli M, Quevedo F, Shukla P (2020) A systematic approach to Kähler moduli stabilisation. JHEP 08:047. https://doi.org/10.1007/JHEP08(2020)047. https://arxiv.org/abs/2005.11329
17. Wess J, Bagger J (1992) Supersymmetry and supergravity. Princeton University Press, Princeton, NJ, USA
18. Gates SJ, Grisaru MT, Rocek M, Siegel W (1983) Superspace or one thousand and one lessons in supersymmetry. Front Phys 58. https://arxiv.org/abs/hep-th/0108200
19. Grana M, Polchinski J (2001) Supersymmetric three form flux perturbations on AdS(5). Phys Rev D 63:026001. https://doi.org/10.1103/PhysRevD.63.026001. https://arxiv.org/abs/hep-th/0009211
20. Giddings SB, Kachru S, Polchinski J (2002) Hierarchies from fluxes in string compactifications. Phys Rev D 66:106006. https://doi.org/10.1103/PhysRevD.66.106006. https://arxiv.org/abs/hep-th/0105097
21. Grana M, Polchinski J (2002) Gauge/gravity duals with holomorphic dilaton. Phys Rev D 65:126005. https://doi.org/10.1103/PhysRevD.65.126005. https://arxiv.org/abs/hep-th/0106014
22. Myers RC (1999) Dielectric branes. JHEP 12:022. https://doi.org/10.1088/1126-6708/1999/12/022. https://arxiv.org/abs/hep-th/9910053
23. Maldacena JM, Nunez C (2001) Supergravity description of field theories on curved manifolds and a no go theorem. Int J Mod Phys A 16:822. https://doi.org/10.1142/S0217751X01003937. https://arxiv.org/abs/hep-th/0007018

24. Hamada Y, Hebecker A, Shiu G, Soler P (2021) Completing the D7-brane local gaugino action. JHEP 11:033. https://doi.org/10.1007/JHEP11(2021)033. https://arxiv.org/abs/2105.11467
25. Kachru S, Kallosh R, Linde AD, Trivedi SP (2003) De Sitter vacua in string theory. Phys Rev D 68:04605. https://doi.org/10.1103/PhysRevD.68.046005. https://arxiv.org/abs/hep-th/0301240
26. Kachru S, Kim M, Mcallister L, Zimet M (2021) de Sitter vacua from ten dimensions. JHEP 12:111. https://doi.org/10.1007/JHEP12(2021)111. https://arxiv.org/abs/1908.04788
27. Polchinski J, Chaudhuri S, Johnson C (1996) Notes on D-branes. https://arxiv.org/abs/hep-th/9602052
28. Johnson CV (2005) D-branes, Cambridge monographs on mathematical physics. Cambridge University Press. https://doi.org/10.1017/CBO9780511606540
29. Becker K, Becker M, Schwarz JH (2006) String theory and M-theory: a modern introduction. Cambridge University Press
30. Green MB, Schwarz JH, Witten E (1988) Superstring theory. Vol. 2: Loop amplitudes, anomalies and phenomenology
31. Klebanov IR, Strassler MJ (2000) Supergravity and a confining gauge theory: duality cascades and chi SB resolution of naked singularities. JHEP 08:052. https://doi.org/10.1088/1126-6708/2000/08/052. https://arxiv.org/abs/hep-th/0007191
32. Burgess CP (2006) Lectures on cosmic inflation and its potential stringy realizations. PoS P 2GC:008. https://doi.org/10.1088/0264-9381/24/21/S04. https://arxiv.org/abs/0708.2865
33. McAllister L, Silverstein E (2008) String cosmology: a review. Gen. Rel. Grav. 40:565. https://doi.org/10.1007/s10714-007-0556-6. https://arxiv.org/abs/0710.2951
34. Baumann D, Dymarsky A, Kachru S, Klebanov IR, McAllister L (2009) Holographic systematics of D-brane inflation. JHEP 03:093. https://doi.org/10.1088/1126-6708/2009/03/093. urlhttps://arxiv.org/abs/0808.2811
35. Pajer E (2008) Inflation at the tip. JCAP 04:031. https://doi.org/10.1088/1475-7516/2008/04/031. https://arxiv.org/abs/0802.2916
36. Avgoustidis A, Zavala I (2009) Warped Wilson line DBI inflation. JCAP 01:045. https://doi.org/10.1088/1475-7516/2009/01/045. https://arxiv.org/abs/0810.5001
37. Taylor TR, Vafa C (2000) R R flux on Calabi-Yau and partial supersymmetry breaking. Phys Lett B: 474:130. https://doi.org/10.1016/S0370-2693(00)00005-8. https://arxiv.org/abs/hep-th/9912152
38. Becker K, Becker M, Haack M, Louis J (2002) Supersymmetry breaking and alpha-prime corrections to flux induced potentials. JHEP 06:060. https://doi.org/10.1088/1126-6708/2002/06/060. https://arxiv.org/abs/hep-th/0204254
39. DeWolfe O, Giddings SB (2003) Scales and hierarchies in warped compactifications and brane worlds. Phys Rev D 67:066008. https://doi.org/10.1103/PhysRevD.67.066008. https://arxiv.org/abs/hep-th/0208123
40. de Wit B, Smit DJ, Hari Dass ND (1987) Residual supersymmetry of compactified D=10 supergravity. Nucl Phys B 283:165. https://doi.org/10.1016/0550-3213(87)90267-7
41. Polchinski J (1995) Dirichlet branes and Ramond-Ramond charges. Phys Rev Lett 75:4724. https://doi.org/10.1103/PhysRevLett.75.4724. https://arxiv.org/abs/hep-th/9510017
42. Denef F (2008) Les Houches lectures on constructing string vacua. Les Houches Series 87:483. https://arxiv.org/abs/0803.1194
43. Cicoli M, Conlon JP, Quevedo F (2008) Systematics of string loop corrections in type IIB Calabi-Yau flux compactifications. JHEP 01:052. https://doi.org/10.1088/1126-6708/2008/01/052. https://arxiv.org/abs/0708.1873
44. Cicoli M, Conlon JP, Quevedo F (2008) General analysis of LARGE volume scenarios with string loop moduli stabilisation. JHEP 10:105. https://doi.org/10.1088/1126-6708/2008/10/105. https://arxiv.org/abs/0805.1029
45. AbdusSalam S, Abel S, Cicoli M, Quevedo F, Shukla P (2020) A systematic approach to Kähler moduli stabilisation. JHEP 08:047. https://doi.org/10.1007/JHEP08(2020)047. https://arxiv.org/abs/2005.11329

46. Cicoli M, Quevedo F, Savelli R, Schachner A, Valandro R (2021) Systematics of the α' expansion in F-theory. JHEP 08:099. https://doi.org/10.1007/JHEP08(2021)099. https://arxiv.org/abs/2106.04592
47. Palti E (2019) The swampland: introduction and review. Fortsch Phys 67:1900037. https://doi.org/10.1002/prop.201900037. https://arxiv.org/abs/1903.06239
48. Smith RC (2019) Notes on the swampland (4): the distance conjecture for arbitrary Calabi-Yau manifolds, the emergence proposal, and the de Sitter conjecture
49. Vafa C (2005) The string landscape and the swampland. https://arxiv.org/abs/hep-th/0509212
50. Ooguri H, Vafa C (2006) On the geometry of the string landscape and the swampland. Nucl Phys B 766:21. https://doi.org/10.1016/j.nuclphysb.2006.10.033. https://arxiv.org/abs/hep-th/0605264
51. Klaewer D, Palti E (2017) Super-Planckian spatial field variations and quantum gravity. JHEP 01:088. https://doi.org/10.1007/JHEP01(2017)088. https://arxiv.org/abs/1610.00010
52. Baume F, Palti E (2016) Backreacted axion field ranges in string theory. JHEP 08:043. https://doi.org/10.1007/JHEP08(2016)043. https://arxiv.org/abs/1602.06517
53. Palti E (2019) The swampland: introduction and review. Fortsch Phys 67:1900037. https://doi.org/10.1002/prop.201900037. https://arxiv.org/abs/1903.06239
54. Valenzuela I (2017) Backreaction issues in axion monodromy and Minkowski 4-forms. JHEP 06:098. https://doi.org/10.1007/JHEP06(2017)098. https://arxiv.org/abs/1611.00394
55. Bielleman S, Ibanez LE, Pedro FG, Valenzuela I, Wieck C (2017) Higgs-otic inflation and moduli stabilization. JHEP 02:073. https://doi.org/10.1007/JHEP02(2017)073. https://arxiv.org/abs/1611.07084
56. Blumenhagen R, Valenzuela I, Wolf F (2017) The swampland conjecture and F-term axion monodromy inflation. JHEP 07:145. https://doi.org/10.1007/JHEP07(2017)145. https://arxiv.org/abs/1703.05776
57. Hebecker A, Henkenjohann P, Witkowski LT (2017) Flat monodromies and a moduli space size conjecture. JHEP 12:033. https://doi.org/10.1007/JHEP12(2017)033. https://arxiv.org/abs/1708.06761
58. Cicoli M, Ciupke D, Mayrhofer C, Shukla P (2018) A geometrical upper bound on the inflaton range. JHEP 05:001. https://doi.org/10.1007/JHEP05(2018)001. https://arxiv.org/abs/1801.05434
59. Font A, Herráez A, Ibáñez LE (2019) The swampland distance conjecture and towers of tensionless branes. JHEP 08:044. https://doi.org/10.1007/JHEP08(2019)044. https://arxiv.org/abs/1904.05379
60. Agrawal P, Obied G, Steinhardt PJ, Vafa C (2018) On the cosmological implications of the string swampland. Phys Lett B 784:271. https://doi.org/10.1016/j.physletb.2018.07.040. https://arxiv.org/abs/1806.09718
61. Denef F, Hebecker A, Wrase T (2018) de Sitter swampland conjecture and the Higgs potential. Phys Rev D 98:086004. https://doi.org/10.1103/PhysRevD.98.086004. https://arxiv.org/abs/1807.06581
62. Garg SK, Krishnan C (2019) Bounds on slow roll and the de Sitter swampland. JHEP 11:075. https://doi.org/10.1007/JHEP11(2019)075. https://arxiv.org/abs/1807.05193
63. Ooguri H, Palti E, Shiu G, Vafa C (2019) Distance and de Sitter conjectures on the swampland. Phys Lett B 788:180. https://doi.org/10.1016/j.physletb.2018.11.018. https://arxiv.org/abs/1810.05506
64. Dine M, Seiberg N (1985) Is the superstring weakly coupled? Phys Lett B 16:299. https://doi.org/10.1016/0370-2693(85)90927-X

Chapter 3
Some Open Questions in Modern Cosmology

Throughout this thesis we will concern ourselves with the consequences of a number of important open questions in modern physics and more particularly, its implications for cosmology. In Chap. 4 we will discuss how to generate a naturally light dark energy field. In Chap. 5, we discuss the implications of the coincidence problem and a possible stringy solution. Naturalness (in the context of cosmology) and the coincidence problem are intimately related to the cosmological constant problem. A problem which is, to a large extent, the underlying motivation for a lot of the research we have undertaken.

In this chapter, we will begin by giving an overview of the naturalness and hierarchy problem in Sect. 3.1. We will then shortly review some problems in cosmology, diving into the well-known cosmological constant problem in Sect. 3.2. Finally, we will discuss the coincidence problem in Sect. 3.3.

3.1 The Hierarchy Problem and Naturalness

In theoretical physics, the original setting for the hierarchy problem is often thought of to be in the context of grand unified theories [1, 2]. To see the hierarchy problem in action, we can take the much simpler example of a scalar ϕ^4 effective theory with some UV cut-off Λ given by the Lagrangian

$$\mathcal{L} = -\frac{1}{2}(\partial_\mu \phi)^2 - \frac{1}{2}m^2\phi^2 - \frac{\lambda}{4!}\phi^4. \tag{3.1}$$

We are interested in calculating the effects of one-loop quantum corrections to the above Lagrangian. The tree-level Feynman diagrams for the theory are given by

F. Cunillera, *Dark Energy*, Springer Theses,
https://doi.org/10.1007/978-3-031-21468-4_3

 $= D^{(2)}$ 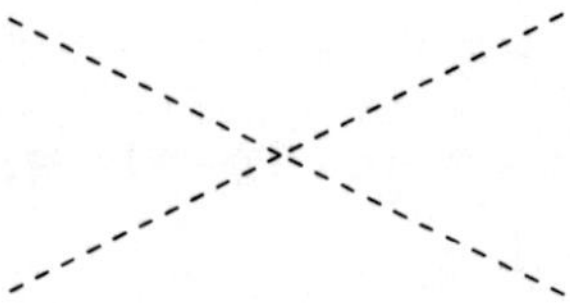 $= D^{(4)}$. (3.2)

By standard calculation techniques [3], the two-point and four-point propagators in Fourier space are given by

$$D^{(2)} = \frac{-i}{p^2 + m^2 + i\varepsilon}, \quad D^{(4)} = -i\lambda, \tag{3.3}$$

and the one loop-diagrams are

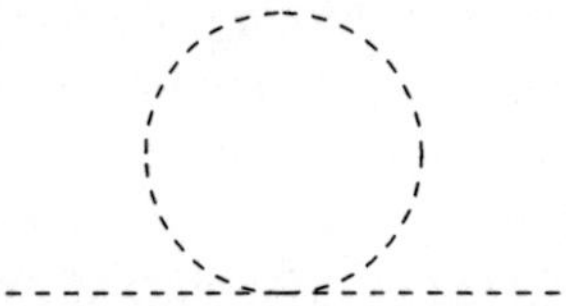 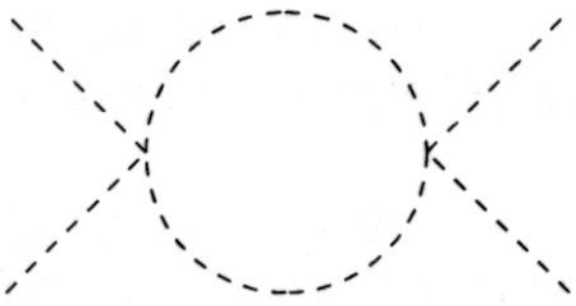 . (3.4)

The introduction of the one-loop diagrams corrects the tree-level expressions such that the integral along a loop carrying internal momenta q^μ is

$$\int_0^\Lambda \frac{d^4q}{(2\pi)^4} \frac{-i}{q^2 + m^2 + i\varepsilon} \frac{-i}{(p-q)^2 + m^2 + i\varepsilon} \propto \ln\left(\frac{\Lambda}{p}\right). \tag{3.5}$$

We can see that in the limit $\Lambda \to \infty$ the integral diverges. More importantly for our discussion, the theory and the physical observables now have knowledge of the cut-off scale Λ. For example, the scattering amplitude for the 2-to-2 particle scattering diagrams is now $i\mathcal{M} \propto -i\lambda + i\lambda^2 \ln\left(\frac{\Lambda^2}{m^2}\right)$. Another related problem is that the self-energy diagram contributes to the two point correlation function with a term

$$\Gamma^{1-\text{loop}}_{(2)} \propto \lambda \int_0^\Lambda \frac{d^4q}{(2\pi)^4} \frac{1}{q^2 + m^2 + i\varepsilon} \propto \lambda \left[\Lambda^2 - m^2 \ln\left(\frac{\Lambda^2}{m^2}\right)\right]. \tag{3.6}$$

The canonical solution to these issues is to introduce counterterms to renormalise the theory. After doing so, the *renormalised* observables of the theory are cut-off independent. For example, introducing the mass counterterm

$$\mathcal{L}_{\delta m} = -\frac{\delta m^2}{2}\phi^2, \quad \delta m^2 = -\frac{\lambda}{32\pi^2}\left[\Lambda^2 - m^2 \ln\left(\frac{\Lambda^2}{m^2}\right)\right], \tag{3.7}$$

turns the two-point correlation function into its renormalised version

$$\bar{\Gamma}_{(2)} = i\left(p^2 + m^2\right), \tag{3.8}$$

to one-loop, with higher orders in loop corrections being systematically reabsorbed into the definition of the counterterms.

This has allowed us to decouple the effective description of the low-energy physics from the cut-off of the theory, and got rid of the UV divergence that appear by sending the cut-off to arbitrarily large values. This will require an infinite number of counterterms but, *a priori*, we have cured the theory from UV divergences. We would like to go one step further now and ask what happens for a theory with a light scalar ϕ and a heavy scalar φ. Let us now consider a theory with two scalar sectors coupled to each other

$$\mathcal{L} = -\frac{1}{2}(\partial_\mu \phi)^2 - \frac{1}{2}m^2\phi^2 - \frac{\lambda_\phi}{4!}\phi^4 - \frac{1}{2}(\partial_\mu \varphi)^2 - \frac{1}{2}M^2\varphi^2 - \frac{\lambda_\varphi}{4!}\varphi^4 - \frac{g}{4}\varphi^2\phi^2, \tag{3.9}$$

here $m \ll M$. Much like in the previous case, one must regularise the theory. After doing so, the tree-level one-to-one scatterings are decoupled from one another since there is no quadratic term mixing ϕ and φ. However, the one-loop diagrams for the light scalar now contain an extra contribution

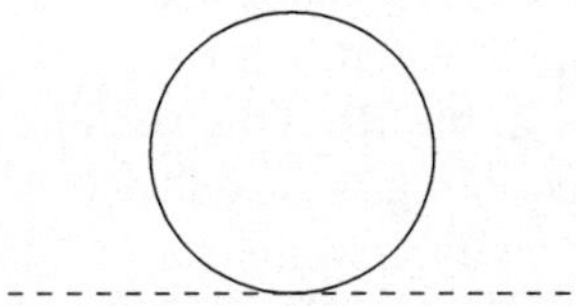

, (3.10)

where the solid line represents the heavy scalar φ. By reproducing the calculations for the renormalised two-point correlator, we find

$$\bar{\Gamma}_{(2)} \sim i\left(p^2 + m^2\right) - \frac{i\lambda_\phi m^2}{32\pi^2} - \frac{igM^2}{32\pi^2}. \tag{3.11}$$

The physics associated to the light sector now depend on the heavy sector, *i.e.* the light and heavy physics only decouple at tree-level. We could try to retune the counterterms to reabsorb the heavy sector dependence. However, by continuing our perturbative expansion to two-loops we would find diagrams of the type

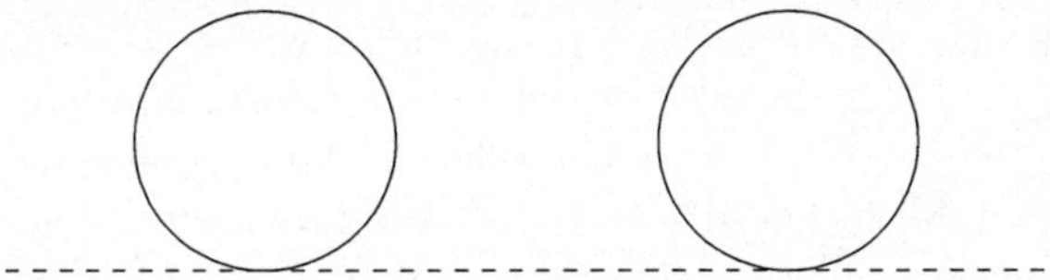

, (3.12)

which would again contribute a term of $\mathcal{O}(gM^2)$ to the two-point correlation function. We would have to fine-tune to the same accuracy at every order in perturbation theory, and we would say that the light scalar is unstable against radiative corrections. This is effectively an infinite fine-tuning and the theory loses all power of predictability. The hierarchy problem can then be restated as: *how can we keep the heavy physics decoupled from the light physics?*

A simple solution would be to argue that g is strictly vanishing. Then the theory with two scalars turns into two theories with one scalar each, analogous to the one in (3.1).This implies the existence of an extra symmetry in the theory. However, consider the case of the Higgs, for example, where the interactions between different sectors are crucial to describe our Universe, then $g \neq 0$ by necessity.

Another possibility is that the symmetry does not fix $g = 0$ but rather introduces new degrees of freedom such that the dependence on the heavy physics drops off from the observables. Of course, this is the idea behind supersymmetry [4] where each component in a supersymmetric multiplet contributes with opposite sign to the n-point correlators, by virtue of supersymmetry pairing fermions to bosons and vice versa. However, no supersymmetric partner to the Standard Model particles has been found at energies $\mathcal{O}(\text{TeV}) \sim 10^{-15} M_{\text{pl}}$, and supersymmetry must then be softly broken at some low-energy scale $E_{SUSY} \gtrsim 10^{-15} M_{\text{pl}}$.

Finally, one could just set $g \ll 1$ by hand such that, even if the two different sectors are coupled, the effects due to heavy physics are subleading to the light sector physics. However, two considerations arise. One is that the usual theoretical physics' lore would tell us that the smallness of g has to be protected by some approximate symmetry in the theory. Otherwise, the RG flow running for g towards arbitrarily high UV physics can spoil the decoupling. A second one is that, in Physics, we do not find arbitrarily tuned constants. Usually, couplings are $\mathcal{O}(1)$; if the couplings are very small there is an underlying reason. This sentiment is the idea behind naturalness. We say that a theory is natural when it contains only couplings which are $\mathcal{O}(1)$. Setting $g \ll 1$ by hand would be unnatural.

More concretely, we introduce 't Hooft's definition of naturalness [5],

Naturalness. *Consider an effective theory valid up to some cut-off* Λ. *Make any free parameters dimensionless by measuring them in units of the cut-off. A theory is natural if the only parameters that are much smaller than unity imply an enhancement of the symmetry group of theory in the limit of those parameters vanishing. Otherwise, the theory is unnatural.*

To exemplify the meaning of an enhanced symmetry, we take our simple one-field ϕ^4 model and follow the discussion of [5]. Fixing $\lambda = 0$ leads to a non-interacting theory, a freely propagating massive particle with a conserved particle number, $\lambda \neq 0$ breaks particle number conservation. Therefore, a theory with $\lambda = 0$ is natural despite having a vanishingly small parameter since $\lambda = 0$ induces a new symmetry in the theory. On the other hand, $m = 0$ does not enhance the symmetries of the theory in any way, and thus fixing $m/\Lambda \ll 1$ by hand would render the theory unnatural. Similarly, in the light-heavy Lagrangian (3.9) setting $g = 0$ does not recover any symmetries, as the total number of particles is still not conserved due to $\lambda_\phi,\ \lambda_\varphi \neq 0$.

As discussed, within the framework of particle physics the hierarchy problem and naturalness issues appear under the guise of the Higgs. The Higgs mass is much smaller than the mass of the top gauge boson, the heaviest Standard Model particle. As mentioned, supersymmetry aims to solve this. Another proposed solution is technicolour [6–8], where the Higgs is a composite particle of two new fermions and the underlying theory in terms of the two fermions is natural. A more in-depth review of the historical developments and philosophy around naturalness and the hierarchy problem can be found in [9].

We can also think of the hierarchy problem and naturalness in more general terms as being a question regarding how to introduce parametrically large scale differences in a theory. For example, the cosmological constant energy scale is given by $E_{cc} \sim \sqrt{\Lambda} \sim 10^{-60} M_{\rm pl}$ whereas top collider physics have an associated energy scale $E_{\rm top} \sim 10^{-16} M_{\rm pl}$ while quantum gravity is often associated with $E_{QG} \sim \mathcal{O}(M_{\rm pl})$ physics. A full theory describing all of these phenomena, and anything in between, would have to span 10^{60} orders of magnitude. If string theory is to describe our Universe, it will certainly have to contain small numbers and whether it can do so in a natural way is non-trivial.

3.2 The Cosmological Constant Problem

In the previous section, we reviewed how light modes in the effective theory can become unstable due to radiative corrections. This statement also affects the cosmological constant, making the theoretical cosmological constant orders of magnitude larger than the observed value for the vacuum energy of our de Sitter Universe. This problem goes by the name of the cosmological constant problem [10], for a modern formulation of this problem, and possible resolutions, we direct the interested reader to [11, 12], as well as references therein.

To see the problem in action, we consider the one-loop vacuum diagrams for the ϕ^4-theory of (3.1) in a flat spacetime

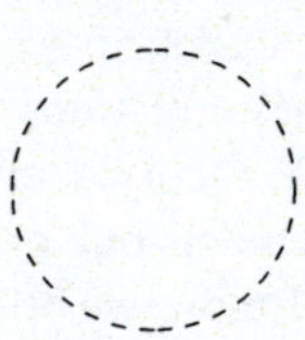

$$= \int \frac{d^4p}{(2\pi)^4} \log\left(p^2 + m^2\right) \propto m^4 \int d^4x \subset -V_{\rm vac} \int d^4x \,, \tag{3.13}$$

where we have renormalised the ε pole, obtained by calculating the loop integral through the usual dimensional renormalisation methods, as this will not play a role in our discussion. In a theory with N particle species, we might expect the vacuum energy to be given by

$$V_{\rm vac} \sim \sum_{0<i\leq N} c_i m_i^4 , \tag{3.14}$$

where c_i are arbitrary coefficients that by naturalness we would expect to be $\mathcal{O}(1)$ and m_i denotes the renormalised mass of the i-th particle species. When gravity kicks in, the vacuum energy will gravitate in the usual way

$$V_{\text{vac}} \int d^4x \rightarrow V_{\text{vac}} \int d^4x \sqrt{-g}, \tag{3.15}$$

and in terms of Feynman diagrams this can be thought of as being given by the sum of vacuum diagrams with an arbitrary number of graviton legs

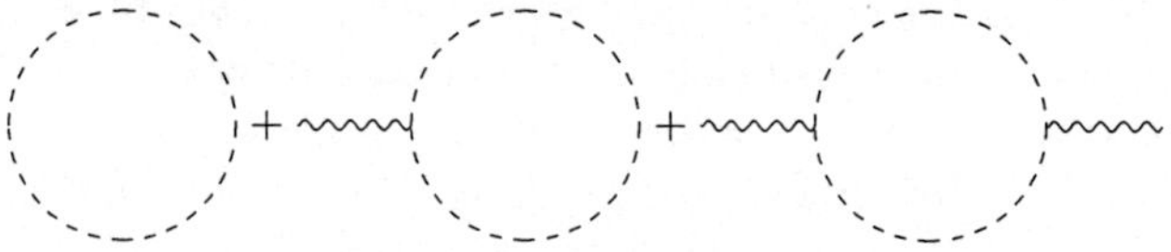

$$+ \ldots \sim -V_{\text{vac}} \int d^4x \sqrt{-g}\,. \tag{3.16}$$

After renormalisation, the one-loop contribution to the vacuum energy for this scalar field will again be of $\mathcal{O}(m^4)$. If we consider the matter sector to be given by the Standard model, we would expect that $V_{\text{vac}} \sim m_{\text{top}}^4 \sim \mathcal{O}(\text{TeV})^4$ which is much heavier than the observed value $V_{\text{vac}}^{\text{observed}} \sim \mathcal{O}(\text{meV})^4$, this constitutes a tuning to an accuracy of 60 orders of magnitude at one-loop. However, we know that the problems do not stop here. Analogously to the discussion of the last section, let us imagine that we do fine-tune the counterterms to allow the theoretical value of the cosmological constant to match the observational one at one-loop. Then, we consider the effects of two-loop diagrams like

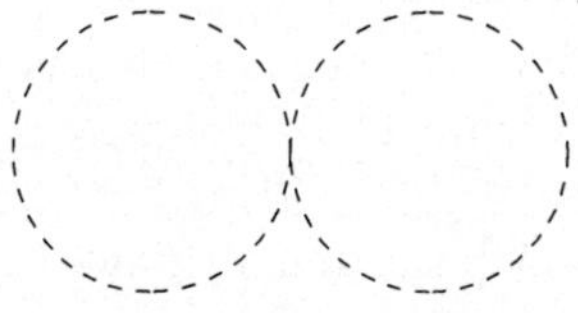

$$= -\lambda \left(\int \frac{d^4p}{(2\pi)^4} \frac{1}{p^2 + m^2 + i\epsilon} \right)^2 \propto -\lambda m^4 \tag{3.17}$$

which contributes to the vacuum energy of the theory at the same order as the one-loop diagrams. This will continue to hold true for higher orders loop diagrams. Again, we find that the cosmological constant is unstable against radiative corrections and that, to match the observations, we must fine-tune the effective field theory at every level in the loop expansion. The cosmological constant problem remains one of the most important and challenging open questions in modern physics. Below, we briefly review some of the proposed resolutions.

3.2.1 The Anthropic Principle

The failure to obtain a microscopic explanation for the smallness of the cosmological constant Λ led researchers to look for a solution in the form of the anthropic principle. The philosophy behind the anthropic principle is best encapsulated in the statement: "Λ should be small enough to allow the Universe to evolve for long enough such that there are scientists that would ask about its smallness". If Λ is much larger in magnitude than its observed value, dark energy domination would start sooner and structure formation could be affected. For concreteness, let us imagine a Universe with $\Lambda \sim \pm M_{\rm pl}^4$. In this Universe, we would not fret about small numbers regarding Λ and thus the cosmological problem would be no more. However, no galaxy formation could have happened and no civilisation could ask questions about the cosmological constant in the first place. On the other hand, if galaxy formation bounds the cosmological constant to small enough values, then the cosmological constant problem can be solved by arguing that a small cosmological constant is the only acceptable value that would sustain intelligent life, which would then go on to worry about its smallness.

In [13], Weinberg builds upon previous anthropic considerations[1] to provide a bound on the available window for the cosmological constant. The anthropic bound on the cosmological constant is given as

$$-10^{-120} M_{\rm pl}^4 < \Lambda < 10^{-118} M_{\rm pl}^4 \,, \tag{3.18}$$

where the upper bound is obtained from demanding that dark energy domination must not begin until sufficiently late times such that structure formation can go through and the lower bound follows from demanding that the Universe does not collapse during matter domination, while stars and galaxies form. Subsequent work has tried to strengthen the bound [15–17]. Work on anthropics has also inspired other resolutions to the cosmological constant problem and the coincidence problem, which we discuss in the following.

Although anthropic considerations might seem anti-climactic in nature, these are by no means trivial statements. A number of underlying assumptions about the true nature of our Universe are at work. For anthropics to solve the cosmological constant problem we require that our Universe can select from different values of the cosmological constant, that at least one of such values is in the observational range and that there exists a dynamical mechanism to allow some regions to attain such value. Since no concrete example of such theories exists, anthropic considerations remain as a proposed resolution.

---

[1] See [14] for a review of the work on anthropics prior to [13].

3.2.2 Sequestering the Vacuum Energy

A qualitatively different approach to solving the cosmological constant problem relies on vacuum energy sequestering [18–23]. In this case, the radiative corrections are reabsorbed in new rigid degrees of freedom. The theory remains locally flat while global dynamics are modified. The mechanism itself is reminiscent of decapitation [24, 25].

Let us consider a semiclassical action of the form

$$S = \int d^4x\sqrt{-g}\left[\frac{1}{2}R - \lambda^4\mathcal{L}_m(\lambda^{-2}g^{\mu\nu}, \Psi) - \Lambda\right] + \sigma\left(\frac{\Lambda}{\lambda^4\mu^4}\right), \tag{3.19}$$

with μ some mass scale and where the classical gravitation sector has been coupled to a quantum matter sector $\mathcal{L}_m$, which we take to contain the Standard model. The global parameter λ is introduced by hand to couple matter Ψ to the cosmological constant Λ, which has been promoted to a global dynamical variable. Finally, σ is a global interaction term, which can be seen to be the dual of a top-form.

The equations of motion for the global variables λ, Λ and the metric are

$$\sigma'\left(\frac{\Lambda}{\lambda^4\mu^4}\right) = \lambda^4\mu^4\int d^4x\sqrt{-g}\,, \tag{3.20}$$

$$\sigma'\left(\frac{\Lambda}{\lambda^4\mu^4}\right) = \frac{\lambda^4\mu^4}{4\Lambda}\int d^4x\sqrt{-g}T^\mu_\mu, \tag{3.21}$$

$$G_{\mu\nu} = -\Lambda g_{\mu\nu} + T_{\mu\nu}, \tag{3.22}$$

where $T_{\mu\nu}$ is the energy momentum tensor for the matter sector with respect to the metric $g_{\mu\nu}$. The last equation is just Einstein's equations, the other two equations allow us to fix the cosmological constant counterterm. By combining the first two equations of motion we find

$$\Lambda = \frac{1}{4}\frac{\int d^4x\sqrt{-g}T^\mu_\mu}{\int d^4x\sqrt{-g}} := \frac{1}{4}\langle T^\mu_\mu\rangle, \tag{3.23}$$

where $\langle T^\mu_\mu\rangle$ is understood as a spacetime average. Furthermore, we write the energy-momentum tensor as $T_{\mu\nu} = -V_{\text{vac}}\,g_{\mu\nu} + \tau_{\mu\nu}$ where $\tau_{\mu\nu}$ describes local excitations and V_{vac} is the vacuum energy coming from Standard model loops. Plugging this expansion for the energy-momentum tensor and (3.23) into (3.22) we find

$$G_{\mu\nu} = \tau_{\mu\nu} - \frac{1}{4}g_{\mu\nu}\langle\tau_{\mu\nu}\rangle. \tag{3.24}$$

meaning that the radiative corrections due to Standard model loops always drop out from the effective field theory. This holds true at any order in loop expansion due to the diffeomorphism invariance of the action and the fact that matter couples universally to λ through the metric $g_{\mu\nu}$, ensuring that at any order in loops the vacuum energy contribution scales like λ^4.

Sequestering resolves the cosmological constant problem at an effective field theory level, but the action has to take a very special form. Without a better understanding of why the effective field theory would have to take that particular form or what kind of UV embedding would yield a low-energy effective action of this kind, we cannot justify sequestering as the one panacea solving all our radiative problems.

3.2.3 Supersymmetry

Another possible resolution could be supersymmetry. After all, supersymmetry came to the rescue when we considered the hierarchy problem for the Standard model. The problem here is that a de Sitter Universe is necessarily non-supersymmetric,[2] meaning that supersymmetry must be (at least) softly broken below some scale E_{SUSY}. In this case, the bosonic-fermionic superpartners have a split in their mass spectrum that goes like

$$m_B^2 - m_F^2 \simeq g_s^2 E_{\text{SUSY}}^2 \,, \tag{3.25}$$

and this ruins the cancellation for the vacuum loop diagrams giving raise to vacuum energy contribution of $\mathcal{O}(g_s^2 E_{\text{SUSY}}^2)$. Since we have not observed supersymmetry in the LHC, we can at most hope to explain why the cosmological constant does not have an M_{pl}^4 energy scale, assuming $E_{\text{SUSY}} < M_{\text{pl}}$, but the coupling problems to the standard model sector remain. It should be noted that one could try to engineer a situation where supersymmetry is broken at two different scales $E_{\text{brane}} \gg E_{\text{bulk}}$, so that the matter sector living in a D3-brane has $E_{\text{SUSY}} = E_{\text{brane}}$ and gravity propagating in the bulk has $E_{\text{SUSY}} = E_{\text{bulk}}$. This type of scenario was already considered in braneworld constructions where the extra dimensions had infinite volume [31, 32] and later in supersymmetric large extra dimensions brane models [33–35]. It remains unclear how to achieve this from a UV perspective. The problem stems from demanding a (nearly) flat four-dimensional solution in the brane. This requires the brane couplings to preserve bulk scale invariance and leads to a fine-tuning in the parameters of the model to achieve the observed curvature scales [36, 37].

3.2.4 The Flux Landscape

To some extend motivated by the anthropic prospects discussed previously, Bousso and Polchinski [38] proposed how we may achieve a small enough cosmological constant through four-form quantisation. Indeed, consider the Einstein-Hilbert action

---

[2] We note that supersymmetry may be realised non-linearly on de Sitter space whenever the symmetry is broken spontaneously, as in [26–30].

minimally coupled to a four-form

$$\mathcal{S} = \int d^4x\sqrt{-g}\left(\frac{1}{2}R - \Lambda - \frac{Z}{48}F_4^2\right), \tag{3.26}$$

where Z is a normalisation constant. The four-form equations of motion fix $F_4 = c\varepsilon_4$, where ε_4 is the four-dimensional Levi-Civita symbol and c is a constant related to the Dirac quantisation condition of the four-form, as we will see below. The effective cosmological constant is the given by [38]

$$\Lambda_{\text{eff}} = \Lambda - \frac{Z}{48}F_4^2 = \Lambda + \frac{Zc^2}{2}. \tag{3.27}$$

If c could take any value, the bare cosmological constant Λ could be cancelled to any degree of accuracy. However, in string theory both the four-form and its dual are quantised and for the dual of the four-form the quantisation condition reads

$$\star\, F_4 = F_0 = \frac{en}{Z}, \quad n \in \mathbb{Z}, \tag{3.28}$$

where e is the charge of the membrane coupled to the four-form. In turn, this fixes our arbitrary constant c to take values in the integer set, in units of $\frac{e}{Z}$. For the effective cosmological constant to be efficiently cancelled by the four-form contribution, we require that

$$\Lambda + \frac{e^2n^2}{2Z} \sim 0 \quad n \sim \frac{\sqrt{2Z\,|\Lambda|}}{e}, \tag{3.29}$$

and that the spacing between different effective cosmological constants, associated to different values of n, is below the observed value $\delta\Lambda \sim \frac{e^2n}{Z} < 10^{-120}$, so that a number of effective cosmological constants can fit observations. All together, we find

$$\sqrt{\frac{|\Lambda|}{Z}}e \lesssim 10^{-120}. \tag{3.30}$$

The problem is that the bare cosmological constant Λ will see Standard model loops, meaning that $\Lambda \sim \mathcal{O}\,(\text{TeV})^4$. This implies that $e \lesssim \sqrt{Z}10^{-108}$, for the action in (3.26) we would expect $Z \sim \mathcal{O}(1)$ which would require $e \lesssim \mathcal{O}(10^{-108})$. One four-form is not enough since this is just exchanging a fine-tuning on the cosmological constant by a slightly smaller fine-tuning on the value of the membrane charge e.

However, in typical compactifications, we expect a large number of these four-forms to be present. In [38], it was also shown that if we have some $\mathcal{O}(100)$ of these four-forms arising from a string compactification, then the individual charges only need to be of $\mathcal{O}(1)$ even if Λ couples to the Planck scale. Then, by anthropic considerations, our Universe is one such that the cosmological constant picked the right value to allow for structure formation. Overall, this argument is based on the idea that flux compactifications generate a very large number of vacua such that even exponentially small cosmological constant may exists in the landscape of theories.

In supergravity models of our Universe, the smallness of the cosmological constant is often related to a balance between an uplift contribution and a tuned tree-level superpotential W_0. In [39, 40], it was argued that statistics of classical flux vacua do indeed agree with the intuition that an exponentially-tuned W_0 is achievable in the landscape such that the cosmological constant can be tuned to a similar degree. For example, in the case of KKLT models which we will discuss in Chap. 7 [41, 42] provide realisations of AdS_4 vacua with exponentially small W_0, fitting observations in magnitude but not in sign as an uplift is still required to obtain dS_4. The idea of using the flux landscape to solve the cosmological constant problem remains attractive but no concrete ten-dimensional realisation of a dS_4 vacua has been achieved yet.

3.3 Coincidence Problem

A related but qualitatively distinct puzzle to the cosmological constant problem is the *cosmological coincidence problem* [43]. This is usually presented as a simple question: why now? Why do we happen to live at a time when the energy density of matter and dark energy are roughly comparable?

Let us unpack the problem in a little more detail. The total age of the universe is estimated to be around 13.7 Gyrs [44]. The early phase of radiation domination represents only a small fraction of that history, giving way to matter after around 51 Kyrs, at a redshift of $z_{eq} \approx 3400$. The transition to dark energy domination occurred much later, after around 3.5 Gyrs, at a redshift of $z_{de} \approx 0.5$. This represents roughly a quarter of the universe's current lifetime [45]. This means that the universe has been matter dominated for a significant fraction of its history, so much so that its density today is still comparable to that of dark energy.

Why is this a problem? The issue is the rate at which dark energy dilutes in comparison to matter, if indeed it dilutes at all. If dark energy is a cosmological constant, as in the standard ΛCDM cosmology, its energy density remains constant while the energy density of matter falls off exponentially quickly once we enter the accelerating phase. If we take the standard scenario at face value, dark energy will dominate indefinitely. It seem implausible that we should find ourselves so close to the dawn of dark energy domination, within a single Hubble time. By way of comparison, we note that planetary orbits are expected to exist for another 10^5 Hubble times [46].

In this sense, the coincidence problem can also be understood from the point of view of naturalness. Indeed, one could fix by hand that the matter-dark energy epoch happens at any *present time*. However, this requires a tremendous fine-tuning of the initial conditions during inflation due to the exponential dilution of matter during the dark energy era.

There have been several proposals that touch upon aspects of this problem (see [45]). These range from anthropic considerations [47–50] to k-essence scenarios where dark energy is triggered by matter-radiation equality, coming to dominate within a few billion years [51–53]. Although the latter go some way to explaining the (limited) duration of the matter dominated epoch, they do not address the main question of why we happen to find ourselves so close to the dawn of dark energy domination. In Chap. 5, we will tackle this problem from the point of view of stringy compactifications and in the remainder of this section we dedicate some time to review these proposed solutions and discuss to what extent these proposal solve the coincidence problem.

We should note that the question of the coincidence problem can also be addressed in apocalyptic fashion by bringing the universe to a rapid conclusion. The basic idea is that dark energy domination begins with acceleration before triggering cosmic Armageddon, within a few efolds. The solution to the coincidence problem follows because the accelerated epoch is cut short - it does not go on indefinitely and endures for a similar time to matter domination. This was shown to occur in phantom cosmologies [50, 54], linear quintessence [50, 55] and sequestering scenarios [18–23, 56–58].

3.3.1 Anthropic Considerations

Following arguments that are similar in spirit to the ones in the last section, in a series of papers [47–49], the authors propose an anthropic resolution to the coincidence problem. Below, we provide a summarised version of the argument flow that led the authors to their anthropic solution:

1. We begin by demanding that the advent of dark energy domination, t_Λ, must happen late enough that it does not destroy structure formation.
2. Assume that most of the carbon-based life appears in our Universe around the time of peak carbon production, t_{carbon}.
3. The main contributors to carbon in the interstellar medium are stars in the few solar mass range [59]. Simulations in [60], showed that peak carbon production happens roughly around the time of star formation and galaxy formation, $t_{\text{carbon}} \sim t_{\text{SFR}} \sim t_G$.
4. Further assume that the appearance of intelligent life is not delayed by more than a fraction of the lifetime of main sequence stars ($\sim$ 5–20 Gyrs), $t_G \sim t_{\text{IL}}$.
5. We take our civilisation as a typical example of carbon-based life and define the "present time" t_0 as the time necessary for a civilisation to become advanced enough to ask about the coincidence problem. On a cosmological timescale, given that our civilisation has been around for only some 10^4 years, so we find $t_G \sim t_{\text{IL}} \sim t_0$.
6. Finally, we have to relate the time of dark energy domination to the time of galaxy formation. In [13], Weinberg took values for the redshift such that galaxy formation is barely possible, giving a lower bound to the advent of dark energy domination. However, in [47], it is argued that if we are truly an average observer

we would expect not to observe the limiting value found by Weinberg but rather one where a sizeable portion of the galaxies have already formed. This is found to be $t_G \sim t_\Lambda$.

Putting everything together we find $t_0 \sim t_G \sim t_\Lambda$ and the coincidence problem seems to be resolved. It is clear that some assumptions must be made about the origin and evolution of intelligent life and these have drawn some criticism [61, 62]. It is typically argued that, although anthropic constraints can select values of the cosmological constant which are close to the observed value, these values necessarily carry large uncertainties due to the lack of understanding of the conditions required for intelligent life to appear.

3.3.2 K-Essence Scenarios

Another proposed resolution to the coincidence problem are the k-essence scenarios originating in [51, 52]. Let us follow the analysis of [53]. We restrict ourselves to models with Lagrangian

$$\mathcal{L} = K(\phi)\tilde{p}(X)\,, \tag{3.31}$$

where $K(\phi) > 0$ and $X := \frac{1}{2}\nabla_\mu\phi\nabla^\mu\phi$. Fixing $K(\phi) := 1/\phi^2$, loosely motivated by the expected perturbative supergravity form of the Kähler metric, and defining the variable $y := X^{-1/2}$ and the $\tilde{p}(X(y)) := g(y)/y$, the equation of state parameter and sound speed are given by

$$w(y) = -\frac{g(y)}{yg'(y)}, \quad c_s^2 = \frac{g(y) - yg'(y)}{y^2 g''(y)}. \tag{3.32}$$

Asking that $w < -1$ and $c_s^2 > 0$ implies that $g(y)$ is a convex and decreasing function of y, but is otherwise left undetermined. We would like the field ϕ to have a stable tracker solution (R) during radiation domination and then reach an attractor solution (K) during k-essence domination with $w \simeq -1$. In principle this would render the k-essence models mostly independent of initial conditions during inflation while attaining a observationally viable behaviour at late times. The question is whether these early tracker to late attractor is a generic behaviour for some family of $\tilde{p}(X)$.

In [53], a dynamical analysis of the models of [51, 52] was conducted. The models themselves are given by

$$\tilde{p}_1(X) := -2.01 + 2\sqrt{1+X} + 3 \times 10^{-17}X^3 - 10^{-24}X^4, \tag{3.33}$$

$$\tilde{p}_2(X) := -2.05 + 2\sqrt{1+f(x)}, \tag{3.34}$$

$$f(X) = X - 10^{-8}X^2 + 10^{-12}X^3 - 10^{-16}X^4 + 10^{-20}X^5 - \frac{10^{-24}}{2^6}X^6. \tag{3.35}$$

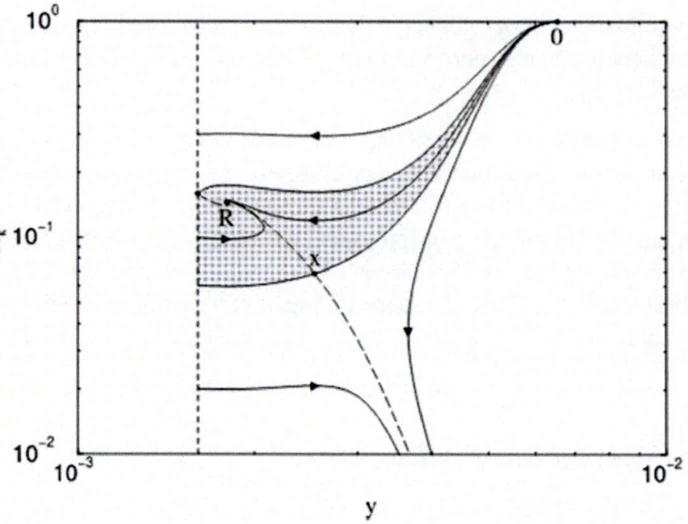

Fig. 3.1 The shaded region represents the basin of attraction for $\tilde{p}_2(X)$ during radiation domination. (R) is the tracker solution, (0) is the origin of the trajectories and (x) is the saddle point defined in [63]. The vertical dashed line describes the point y_c, where $g''(y_c) = 0$ and the sound speed diverges. Figure taken from [53]

Since we are most interested in studying whether the fine-tuning of initial conditions is alleviated in any way, we concentrate on the behaviour of these models during radiation domination and whether the tracker solution exists for a wide range of initial conditions. The dynamical analysis showed that although some basin of attraction exists towards the radiation tracker (R), see for example the phase diagram in Fig. 3.1, the basin is quite limited for these models. For most initial conditions, the model does not reach the tracker solution but rather rapidly evolves towards a period of early k-essence domination or a singularity given by a divergence of the sound speed in the theory.

On top of this, it is clear that the models given in (3.33) and (3.34) have been engineered by hand to contain these tracker solution in the first place. So we are exchanging some slightly less fine-tuned initial conditions for highly contrived models. If the subset of initial conditions turned out to be physically important, then this exchange might be beneficial. However, without any prior or reason to believe that these regions of the parameter space would be more desirable, we find no clear benefits to using the k-essence models in (3.33) and (3.34). Therefore, although an interestingly physical mechanism, k-essence does not constitute a resolution of the coincidence problem insofar as a large tuning of initial conditions seems to still be required.

In this chapter, we have reviewed some open questions of cosmology that will become important in the remainder of the thesis. The question on naturalness was the driving force behind our idea of developing a dark energy model involving the clockwork mechanism, which we describe in Chap. 4. The coincidence problem will be the centrepiece of the discussion in Chap. 5, where we propose how string compactifications might play a role in the resolution to the problem. Finally, the hierarchy problem and the microscopic nature of dark energy will play a key role in the goals and motivations for the second part of the thesis.

References

1. Gildener E (1976) Gauge symmetry hierarchies. Phys Rev D 14:1667. https://doi.org/10.1103/PhysRevD.14.1667
2. Weinberg S (1979) Gauge hierarchies. Phys Lett B 82:387. https://doi.org/10.1016/0370-2693(79)90248-X
3. Peskin ME, Schroeder DV (1995) An Introduction to quantum field theory. Addison-Wesley, Reading, USA
4. Dimopoulos S, Georgi H (1981) Softly broken supersymmetry and SU(5). Nucl Phys B 193:150. https://doi.org/10.1016/0550-3213(81)90522-8
5. 'tHooft G (1980) Naturalness, chiral symmetry, and spontaneous chiral symmetry breaking. NATO Sci Ser B 59:135. https://doi.org/10.1007/978-1-4684-7571-5_9
6. Susskind L (1979) Dynamics of spontaneous symmetry breaking in the Weinberg-Salam theory. Phys Rev D 20:2619. https://doi.org/10.1103/PhysRevD.20.2619
7. Dimopoulos S, Susskind L, Raby S (1980) Technicolour. AIP Conf Proc 59:407. https://doi.org/10.1063/1.32174
8. Farhi E, Susskind L (1979) A technicolored G.U.T. Phys Rev D 20:3404. https://doi.org/10.1103/PhysRevD.20.3404
9. Giudice GF (2008) Naturally speaking: the naturalness criterion and physics at the LHC. https://arxiv.org/abs/0801.2562
10. Weinberg S (1989) The cosmological constant problem. Rev Mod Phys 61:1. https://doi.org/10.1103/RevModPhys.61.1
11. Burgess CP (2015) The cosmological constant problem: why it's hard to get dark energy from micro-physics. In: 100e Ecole d'Ete de Physique: post-Planck cosmology, pp 149–197. https://doi.org/10.1093/acprof:oso/9780198728856.003.0004. https://arxiv.org/abs/1309.4133
12. Padilla A (2015) Lectures on the cosmological constant problem. https://arxiv.org/abs/1502.05296
13. Weinberg S (1987) Anthropic bound on the cosmological constant. Phys Rev Lett 59:2607. https://doi.org/10.1103/PhysRevLett.59.2607
14. Barrow JD, Tipler FJ (1988) The Anthropic cosmological principle. Oxford University Press, Oxford
15. Weinberg S (1996) Theories of the cosmological constant. In: Conference on critical dialogs in cosmology, pp 195–203. https://arxiv.org/abs/astro-ph/9610044
16. Martel H, Shapiro PR, Weinberg S (1998) Likely values of the cosmological constant. Astrophys J 492:29. https://doi.org/10.1086/305016. https://arxiv.org/abs/astro-ph/9701099
17. Garriga J, Vilenkin A (2000) On likely values of the cosmological constant. Phys Rev D 61:083502. https://doi.org/10.1103/PhysRevD.61.083502. https://arxiv.org/abs/astro-ph/9908115
18. Kaloper N, Padilla A (2014) Sequestering the standard model vacuum energy. Phys Rev Lett 112:091304. https://doi.org/10.1103/PhysRevLett.112.091304. https://arxiv.org/abs/1309.6562
19. Kaloper N, Padilla A (2014) Vacuum energy sequestering: the framework and its cosmological consequences. Phys Rev D 90:084023. https://arxiv.org/abs/1406.0711
20. Kaloper N, Padilla A (2015) Sequestration of vacuum energy and the end of the universe. Phys Rev Lett 114:101302. https://doi.org/10.1103/PhysRevLett.114.101302. https://arxiv.org/abs/1409.7073

21. Kaloper N, Padilla A, Stefanyszyn D, Zahariade G (2016) Manifestly local theory of vacuum energy sequestering. Phys Rev Lett 116:051302. https://doi.org/10.1103/PhysRevLett.116.051302. https://arxiv.org/abs/1505.01492
22. Kaloper N, Padilla A, Stefanyszyn D (2016) Sequestering effects on and of vacuum decay. Phys Rev D 94:025022. https://doi.org/10.1103/PhysRevD.94.025022. https://arxiv.org/abs/1604.04000
23. Kaloper N, Padilla A (2017) Vacuum energy sequestering and graviton loops. Phys Rev Lett 118:061303. https://doi.org/10.1103/PhysRevLett.118.061303. https://arxiv.org/abs/1606.04958
24. Adams A, McGreevy J, Silverstein E (2002) Decapitating tadpoles. https://arxiv.org/abs/hep-th/0209226
25. Niedermann F, Padilla A (2017) Gravitational mechanisms to self-tune the cosmological constant: obstructions and ways forward. Phys Rev Lett 119:251306. https://doi.org/10.1103/PhysRevLett.119.251306. https://arxiv.org/abs/1706.04778
26. Bergshoeff EY, Freedman DZ, Kallosh R, Van Proeyen A (2015) Pure de Sitter supergravity. Phys Rev D 92:085040. https://doi.org/10.1103/PhysRevD.93.069901. https://arxiv.org/abs/1507.08264
27. Dudas E, Ferrara S, Kehagias A, Sagnotti A (2015) Properties of nilpotent supergravity. JHEP 09:217. https://doi.org/10.1007/JHEP09(2015)217. https://arxiv.org/abs/1507.07842
28. Bandos I, Martucci L, Sorokin D, Tonin M (2016) Brane induced supersymmetry breaking and de Sitter supergravity. JHEP 02:080. https://doi.org/10.1007/JHEP02(2016)080. https://arxiv.org/abs/1511.03024
29. Nagy S, Padilla A, Zavala I (2020) The super-Stückelberg procedure and dS in pure supergravity. Proc Roy Soc Lond A 476:20200035. https://doi.org/10.1098/rspa.2020.0035. https://arxiv.org/abs/1910.14349
30. Bansal S, Nagy S, Padilla A, Zavala I (2021) Unimodular vs nilpotent superfield approach to pure dS supergravity. JHEP 01:146. https://doi.org/10.1007/JHEP01(2021)146. https://arxiv.org/abs/2010.13758
31. Dvali GR, Gabadadze G, Porrati M (2000) Metastable gravitons and infinite volume extra dimensions. Phys Lett B 484:112. https://doi.org/10.1016/S0370-2693(00)00631-6. https://arxiv.org/abs/hep-th/0002190
32. Witten E (2000) The cosmological constant from the viewpoint of string theory. In: 4th international symposium on sources and detection of dark matter in the universe (DM 2000), pp 27–36. https://arxiv.org/abs/hep-ph/0002297
33. Aghababaie Y, Burgess CP, Parameswaran SL, Quevedo F (2004) Towards a naturally small cosmological constant from branes in 6-D supergravity. Nucl Phys B 680:389. https://doi.org/10.1016/j.nuclphysb.2003.12.015. https://arxiv.org/abs/hep-th/0304256
34. Burgess CP, van Nierop L (2013) Technically natural cosmological constant from supersymmetric 6D brane backreaction. Phys Dark Univ 2:1. https://doi.org/10.1016/j.dark.2012.10.001. https://arxiv.org/abs/1108.0345
35. Burgess CP, van Nierop L (2011) Large dimensions and small curvatures from supersymmetric brane back-reaction. JHEP 04:078. https://doi.org/10.1007/JHEP04(2011)078. https://arxiv.org/abs/1101.0152
36. Niedermann F, Schneider S (2016) Fine-tuning with brane-localized flux in 6D supergravity. JHEP 02:025. https://doi.org/10.1007/JHEP02(2016)025. https://arxiv.org/abs/1508.01124
37. Niedermann F, Schneider R (2016) SLED phenomenology: curvature vs. volume. JHEP 03:130. https://doi.org/10.1007/JHEP03(2016)130. https://arxiv.org/abs/1512.03800
38. Bousso R, Polchinski J (2000) Quantization of four form fluxes and dynamical neutralization of the cosmological constant. JHEP 06:006. https://doi.org/10.1088/1126-6708/2000/06/006. https://arxiv.org/abs/hep-th/0004134
39. Ashok S, Douglas MR (2004) Counting flux vacua. JHEP 01:060. https://doi.org/10.1088/1126-6708/2004/01/060. https://arxiv.org/abs/hep-th/0307049
40. Denef F, Douglas MR (2004) Distributions of flux vacua. JHEP 05:072. https://doi.org/10.1088/1126-6708/2004/05/072. https://arxiv.org/abs/hep-th/0404116

41. Demirtas M, Kim M, McAllister L, Moritz J, Rios-Tascon A (2022) Exponentially small cosmological constant in string theory. Phys Rev Lett 128:011602. https://doi.org/10.1103/PhysRevLett.128.011602. https://arxiv.org/abs/2107.09065
42. Demirtas M, Kim M, McAllister L, Moritz J, Rios-Tasco A (2021) Small cosmological constants in string theory. JHEP 12:136. https://doi.org/10.1007/JHEP12(2021)136. https://arxiv.org/abs/2107.09064
43. Zlatev I, Wang LM, Steinhardt PJ (1999) Quintessence, cosmic coincidence, and the cosmological constant. Phys Rev Lett 82:896. https://doi.org/10.1103/PhysRevLett.82.896. https://arxiv.org/abs/astro-ph/9807002
44. PLANCK Collaboration (2020) Planck 2018 results. VI. Cosmological parameters. Astron Astrophys 641:A6. https://doi.org/10.1051/0004-6361/201833910. https://arxiv.org/abs/1807.06209
45. Velten HES, vom Marttens RF, Zimdahl W (2014) Aspects of the cosmological "coincidence problem". Eur Phys J C 74:3160. https://doi.org/10.1140/epjc/s10052-014-3160-4. https://arxiv.org/abs/1410.2509
46. Adams FC, Laughlin G (1997) A dying universe: the long term fate and evolution of astrophysical objects. Rev Mod Phys 69:337. https://doi.org/10.1103/RevModPhys.69.337. https://arxiv.org/abs/astro-ph/9701131
47. Garriga J, Livio M, Vilenkin A (2000) The cosmological constant and the time of its dominance. Phys Rev D 61:023503. https://doi.org/10.1103/PhysRevD.61.023503. https://arxiv.org/abs/astro-ph/9906210
48. Weinberg S (2000) The cosmological constant problems. In: 4th international symposium on sources and detection of dark matter in the universe (DM 2000), pp 18–26. https://arxiv.org/abs/astro-ph/0005265
49. Bludman SA (2000) Vacuum energy: if not now, then when? Nucl Phys A 663:865. https://doi.org/10.1016/S0375-9474(99)00733-2. https://arxiv.org/abs/astro-ph/9907168
50. Barreira A, Avelino PP (2011) Anthropic versus cosmological solutions to the coincidence problem. Phys Rev D 83:103001. https://doi.org/10.1103/PhysRevD.83.103001. https://arxiv.org/abs/1103.2401
51. Armendariz-Picon C, Mukhanov VF, Steinhardt PJ (2000) A dynamical solution to the problem of a small cosmological constant and late time cosmic acceleration. Phys Rev Lett 85:4438. https://doi.org/10.1103/PhysRevLett.85.4438. https://arxiv.org/abs/astro-ph/0004134
52. Armendariz-Picon C, Mukhanov VF, Steinhardt PJ (2001) Essentials of k essence. Phys Rev D 63:103510. https://doi.org/10.1103/PhysRevD.63.103510. https://arxiv.org/abs/astro-ph/0006373
53. Malquarti M, Copeland EJ, Liddle AR (2003) K-essence and the coincidence problem. Phys Rev D 68:023512. https://doi.org/10.1103/PhysRevD.68.023512. https://arxiv.org/abs/astro-ph/0304277
54. Scherrer RJ (2005) Phantom dark energy, cosmic doomsday, and the coincidence problem. Phys Rev D 71:063519. https://doi.org/10.1103/PhysRevD.71.063519. https://arxiv.org/abs/astro-ph/0410508
55. Avelino PP (2005) The coincidence problem in linear dark energy models. Phys Lett B 611:15. https://doi.org/10.1016/j.physletb.2005.02.037. https://arxiv.org/abs/astro-ph/0411033
56. D'Amico G, Kaloper N, Padilla A, Stefanyszyn D, Westphal A, Zahariade G (2017) An étude on global vacuum energy sequester. JHEP 09:074. https://doi.org/10.1007/JHEP09(2017)074. https://arxiv.org/abs/1705.08950
57. Niedermann F, Padilla A (2017) Gravitational mechanisms to self-tune the cosmological constant: obstructions and ways forward. Phys Rev Lett 119:251306. https://doi.org/10.1103/PhysRevLett.119.251306. https://arxiv.org/abs/1706.04778
58. Coltman B, Li Y, Padilla A (2019) Cosmological consequences of Omnia Sequestra. JCAP 06:017. https://doi.org/10.1088/1475-7516/2019/06/017. https://arxiv.org/abs/1903.02829
59. Iben I, Renzini A (1983) Asymptotic giant branch evolution and beyond. Ann Rev Astron Astrophys 21:271. https://doi.org/10.1146/annurev.aa.21.090183.001415

60. Livio M (1999) How rare are extraterrestrial civilizations and when did they emerge? Astrophys J 511:429. https://doi.org/10.1086/306668. https://arxiv.org/abs/astro-ph/9808237
61. Graesser ML, Hsu SDH, Jenkins A, Wise MB (2004) Anthropic distribution for cosmological constant and primordial density perturbations. Phys Lett B 600:15. https://doi.org/10.1016/j.physletb.2004.08.061. https://arxiv.org/abs/hep-th/0407174
62. Aguirre A (2001) The cold big bang cosmology as a counter example to several anthropic arguments. Phys Rev D 64:083508. https://doi.org/10.1103/PhysRevD.64.083508. https://arxiv.org/abs/astro-ph/0106143astro-ph/0106143
63. Armendariz-Picon C, Mukhanov VF, Steinhardt PJ (2001) Essentials of k essence. Phys Rev D 63:103510. https://doi.org/10.1103/PhysRevD.63.103510. https://arxiv.org/abs/astro-ph/0006373

Chapter 4
A Naturally Light Dark Energy Model

In [1], we proposed a model of dark energy driven by a pseudo-scalar field whose super-light mass emerges naturally from a simple microscopic theory with uniquely high scale couplings. The model is a dark energy avatar of the pion, its low mass emerging from the spontaneous breaking of a weakly broken symmetry. It consists of $\mathcal{O}(100)$ pseudo scalar fields (axions) with non-trivial mass mixing, one of which has a bilinear mixing with a four-form field strength. All mass scales in the theory are assumed to lie at, or close to, the Planck scale. The motivation for this model arises as the marriage between two axion EFTs and we will begin by introducing these models: the clockwork mechanism [2, 3] in Sect. 4.1 and the Dvali-Kaloper-Sorbo model [4–7] in Sect. 4.2.

In Sect. 4.3 we will present our model in full and further motivate its UV completion in terms of a type IIA supergravity motivated toy model in Sect. 4.4. We will then proceed to scrutinise the type IIA embedding in more detail, being able to write down a no-go theorem for the clockwork mechanism at the level of the perturbative type IIA action with fluxes in Sect. 4.5. Finally, in Sect. 4.6 we perform the dimensional deconstruction of the discrete model, and study a family of braneworld configurations showing how the braneworld observer measures a low scale of dark energy on account of the warping of the extra dimension.

4.1 Clockwork Mechanism

The underlying theory that we present in this section is based on the works [2, 3], where the authors present a mechanism to obtain trans-Planckian axion decay constants from a theory with $\mathcal{O}(M_{\rm pl})$ mass parameters. These constructions are intimately related to the problem of obtaining viable natural inflation in string theory. In natural inflation, the large field excursions of the inflation are protected from quantum corrections by virtue of the inflaton being an axion with a discrete shift symmetry. In this way, even if the inflaton range is super-Planckian, the axion range remains sub-Planckian as long as its decay constant is super-Planckian. However,

F. Cunillera, *Dark Energy*, Springer Theses,
https://doi.org/10.1007/978-3-031-21468-4_4

it was pointed out that trans-Planckian decay constant are not easy to engineer in supergravity theories whilst keeping perturbative control [8]. The Kim-Nilles-Peloso (KNP) mechanism [9] provided a way to circumvent this issue by considering two-axion models with sub-Planckian decay constants, in such a way that there exists a linear combination of the axions with super-Planckian decay constant. From the point of view of the EFT, there exists a field with an effectively super-Planckian decay constant that could prove to be a good candidate for the inflaton. However, considerations related to strong forms of the weak gravity conjecture [10–13] put some stringent bounds on the KNP mechanism.

In [14], the authors go one step further and generalise the works [2, 3] to show how to generate parametrically low energy scales from a theory with only Planckian-scale parameters. To do so, they will require the introduction of a technically natural parameter (in the sense of 't Hooft [15]) $\sigma := M^2/f^2 \ll 1$, that represents the scale separation between spontaneously broken physics at a scale f and softly broken symmetries at a scale M. In [14], the clockwork mechanism is generalised to scalars, vectors and tensor fields. For our discussion, we will concentrate on the scalar clockwork.

Let us consider a theory with a global symmetry $\mathcal{G} = U(1)^{N+1}$ spontaneously broken at a scale f. We will take $f \sim \mathcal{O}(M_{\text{pl}})$ but the mechanism generalises straightforwardly to any other scale, where f would represent the cut-off of the would-be effective field theory. Below the symmetry breaking scale, the theory contains $N+1$ Goldstone bosons ϕ_j which can be packaged conveniently as

$$U_j = e^{i\phi_j/f}\,, \qquad j = 0, \ldots, N\,, \tag{4.1}$$

which transform by a phase under the corresponding $U(1)_j$ generator. We also softly break $\mathcal{G}$ by introducing N mass parameters M_i, with $i = 0, \ldots, N-1$. These can be regarded as N spurion fields taking non-zero vacuum expectations values due to their charges under the $U(1)_i$

$$Q_i[M_k^2] = \delta_{i\,k} - q\delta_{i\,k+1}\,. \tag{4.2}$$

Taking $q > 1$, the symmetry breaking will be soft as long as $\sigma = M/f \ll 1$.

The leading terms of the effective Lagrangian in terms of the true scalar and the N pseudo-scalar are

$$\mathcal{L} = -\frac{1}{2}\left[\sum_{i=0}^{N}(\partial\phi_i)^2 + M^2\sum_{i=0}^{N-1}(\phi_i - q\phi_{i+1})^2 + \mathcal{O}(\phi^4)\right], \tag{4.3}$$

where we can rewrite the potential in terms of a mass matrix

$$V(\phi) = \frac{1}{2}M^2\sum_{i=0}^{N-1}(\phi_i - q\phi_{i+1})^2 + \mathcal{O}(\phi^4) = \frac{1}{2}\sum_{i,j=0}^{N}\phi_i\mathcal{M}_{ij}^2\phi_j + \mathcal{O}(\phi^4)\,, \tag{4.4}$$

$$\mathcal{M}^2_{ij} = M^2 \begin{bmatrix} 1 & -q & 0 & \cdots & & & 0 \\ -q & 1+q^2 & -q & & & & \\ 0 & -q & 1+q^2 & \ddots & \ddots & & \vdots \\ & & \ddots & \ddots & \ddots & & \\ \vdots & & \ddots & \ddots & 1+q^2 & -q & 0 \\ & & & & -q & 1+q^2 & -q \\ 0 & & \cdots & & 0 & -q & q^2 \end{bmatrix} . \tag{4.5}$$

Diagonalising the mass matrix, we find for some orthogonal matrix $\mathcal{O}$

$$\tilde{\mathcal{M}}^2_{ij} = \mathcal{O}^T_{ik}\mathcal{M}^2_{kl}\mathcal{O}_{lj} = \text{diag}\left(m_0^2, ..., m_N^2\right) , \tag{4.6}$$

$$m_0^2 = 0\,, \quad m_k^2 = \left[q^2 + 1 - 2q\cos\left(\frac{k\pi}{N+1}\right)\right] M^2 . \tag{4.7}$$

The soft symmetry breaking has broken the axion shift symmetry, providing a mass for N pseudo-scalars. The mass spectrum of the theory contains a massless mode, related to the leftover spontaneously broken $U(1)$, and N massive modes. In the large N limit, the massive modes span masses

$$m_1 \simeq (q-1)M\,, \qquad m_N \simeq (q+1)M = m_1\left(1 + \frac{2}{q-1}\right) , \tag{4.8}$$

and the mass splitting $\delta m_k = m_{k+1} - m_k$ is given by

$$\frac{\delta m_k}{m_k} \simeq \frac{q\pi}{N\left[q^2 + 1 - 2q\cos\left(\frac{k\pi}{N+1}\right)\right]} \sin\left(\frac{k\pi}{N+1}\right) . \tag{4.9}$$

Therefore, for a large number of fields (or "gears" in the language of [14]), the N massive states enter the theory at a mass scale $m_i \sim \mathcal{O}(M)$, with gaps $\delta m_i \sim \mathcal{O}(M/N)$. Furthermore, rewriting the theory in term of the mass eigenstates in the diagonal mass matrix basis $\pi_i = \sum_{j=0}^N \mathcal{O}^T_{ij}\phi_i$, we crucially find that $\mathcal{O}_{0j} \propto q^{-j}$. This means that if an operator couples to the *N-th* gear, the massless mode will feel its interaction suppressed by q^N and for large N this can become an efficient way to "protect" the massless mode against radiative corrections. We will see this explicitly in Sect. 4.3 once we couple the Dvali-Kaloper-Sorbo model to the *N-th* site of the clockwork chain.

4.2 Dvali-Kaloper-Sorbo Four-Form Model

Inspired by a long history of research on applications of membranes to gravity (see for example [16–18]) arises the second element of our model. The Dvali-Kaloper-Sorbo model (DKS) [4–7] has been proposed as an effective theory of axion monodromy in string theory [19, 20]. However, the true string embedding of the DKS model remains unclear [21]. Nonetheless, DKS, as an EFT, can provide an interesting description of single-field axion inflation since it avoids some of the issues related with super-Planckian excursions, in a similar vein to the KNP mechanism.

The idea is to introduce a bilinear mixing between the axion and a four-form field strength. The theory admits a dual description in terms of a massive pseudoscalar and the magnetic flux of the four-form. The latter is locally constant in spacetime although it can jump between quantised values across a three dimensional membrane. The effective inflaton is then a gauge invariant combination of the axion and the magnetic flux—its large field values are obtained through the flux, which in turn may be identified with macroscopic quantities characterizing the system rather than high scale excitations of the inflaton field [22]. Small deformations of these models can also give rise to an emergent mechanism for screening the effects of vacuum energy at large scales [23–25].

The starting point for the DKS model is a massive $U(1)$ gauge theory given by [7]

$$\mathcal{L} = -\frac{1}{48}F^2 - \frac{\mu^2}{12}\left(A_{\mu\nu\alpha} - h_{\mu\nu\alpha}\right)^2 + \frac{\mu}{6}\phi\frac{\epsilon^{\mu\nu\alpha\beta}}{\sqrt{-g}}\partial_\mu h_{\nu\alpha\beta}\,, \tag{4.10}$$

here $F_{\mu\nu\alpha\beta} = 4\partial_{[\mu}A_{\nu\alpha\beta]}$ is the four-form field strength, $\epsilon^{\mu\nu\rho\sigma}$ is the totally antisymmetric Levi-Civita symbol defined such that $\epsilon^{0123} = 1$, with indices raised and lowered with respect to the metric $g_{\mu\nu}$, and ϕ is an axion with periodicity $\phi := \phi + 2\pi f$, with f left to be determined. The equation of motion for ϕ yields $h = db$, with $b_{\mu\nu}$ the Stueckelberg field invariant under $b \to b + d\lambda$. Integrating out the Stueckelberg through its equation of motion yields

$$\mathcal{L} = -\frac{1}{48}F^2 - \frac{1}{2}(\partial\phi)^2 + \frac{\mu}{24}\frac{\epsilon^{\mu\nu\alpha\beta}}{\sqrt{-g}}\phi F_{\mu\nu\alpha\beta}\,. \tag{4.11}$$

We can eliminate the four-form by introducing a Lagrange multiplier Q that fixes the Bianchi identity for the four-form on-shell. This allows us to rewrite the theory in terms of the pseudo-scalar

$$\mathcal{L} = -\frac{1}{2}(\partial\phi)^2 - \frac{\mu^2}{2}\left(\phi + \frac{Q}{\mu}\right)^2 - \frac{Q}{6}\frac{\epsilon^{\mu\nu\alpha\beta}}{\sqrt{-g}}\partial_\mu A_{\nu\alpha\beta}\,, \tag{4.12}$$

which is invariant under

$$\phi \to \phi + 2\pi f\,, \qquad Q \to Q - 2\pi q\,, \tag{4.13}$$

where $f = q/\mu$.

In the DKS model, one identifies the effective inflaton with $\varphi = \phi + Q/\mu$. We can see that, the inflaton gets a mass from its mixing with the four-form field strength, and the stronger the coupling μ the larger the effective mass. It is this mechanism for generating a mass that we exploit in our model of dark energy. By coupling one end of our clockwork to the four-form we guarantee that the mixing with the zero mode of the clockwork is exponentially small and as a result, a very small mass is generated in the low energy effective theory.

4.3 Obtaining a Naturally Light Dark Energy Scale

We begin with a chain of $N + 1$ pseudo-scalar fields $\phi_0, ..., \phi_N$, all characterized by a single ultraviolet mass scale M and a nearest neighbour interaction with strength q [2, 3]. The mass M is assumed to lie at or close to the Planck scale. We further assume that one end of the chain is coupled to a four-form field strength as in the Dvali-Kaloper-Sorbo model. The combined set up is described by the following Lagrangian density:

$$\mathcal{L} = -\frac{1}{2}\left[\sum_{i=0}^{N}(\partial\phi_i)^2 + M^2\sum_{i=0}^{N-1}(\phi_i - q\phi_{i+1})^2\right] + \frac{\mu}{24}\phi_N\frac{\epsilon^{\mu\nu\alpha\beta}}{\sqrt{-g}}F_{\mu\nu\alpha\beta} - \frac{1}{48}F_{\mu\nu\alpha\beta}F^{\mu\nu\alpha\beta}\ . \tag{4.14}$$

In the first line we recognise the clockwork model of [2, 3]. In principle we could allow for site dependent masses $M_i \sim M$ and mixing strengths $q_i \sim q$, although for simplicity we take them all to be equal. The coupling q is dimensionless and assumed to be greater than 1, but it remains of order unity. The second line of Eq. (4.14) contains the DKS model for the N-th site in the chain, in the form of Eq. (4.11) and where we have identified $\phi_N = \phi$. The bilinear mixing between the axion and the four-form reveals another mass scale, $\mu \sim M$, which we also assume to be given by the characteristic ultraviolet scale of the theory. The gravitational sector of the theory is assumed to be described by Einstein gravity, although we will not need to include that in our discussion.

The original clockwork Lagrangian gives rise to a massless Goldstone pseudo-scalar [2, 3, 14], a consequence of the non-linearly realised shift symmetry $\phi_i \to \phi_i + cq^{-i}$ for arbitrary c. This symmetry remains perturbatively unbroken through the mixing with the four-form, although in the presence of charged membranes, non-perturbative effects break the continuous symmetry down to a discrete subgroup [26, 27]. However, the symmetry is only *mildly* broken because ϕ_N has exponentially suppressed overlap with the zero mode. We thus expect that the zero mode acquires a mass, but that the latter remains tiny. In order to see this explicitly it is convenient to integrate out the four-form and pass to a dual description in which the four-form mixing generates a new mass term for the last axion in the chain. This can be done in a straightforward manner by adding a Lagrange multiplier term of the form $\frac{1}{24}Q\frac{\epsilon^{\mu\nu\alpha\beta}}{\sqrt{-g}}(F_{\mu\nu\alpha\beta} - 4\partial_{[\mu}A_{\nu\alpha\beta]})$ then eliminating F using its algebraic equation of motion[1] [6]. This yields a dual theory described by the following Lagrangian:

$$\mathcal{L} = -\frac{1}{2}\left[\sum_{i=0}^{N}(\partial\phi_i)^2 + M^2\sum_{i=0}^{N-1}(\phi_i - q\phi_{i+1})^2\right] \\ -\frac{1}{2}(Q+\mu\phi_N)^2 - \frac{Q}{6}\frac{\epsilon^{\mu\nu\alpha\beta}}{\sqrt{-g}}\partial_{[\mu}A_{\nu\alpha\beta]}\,. \tag{4.15}$$

The Lagrange multiplier Q is fixed to be constant by the variation of the three-form. If the latter is coupled to membrane charges, then Q is quantised in units of the membrane charge, e, as in $\langle Q\rangle = 2\pi Ne$ for integer values of N [26, 27]. This quantisation condition is compatible with the unbroken symmetry transformations, which take the form $\phi_i \to \phi_i + 2\pi n\frac{e}{\mu}q^{N-i}$, $Q \to Q + 2\pi ne$ for integer values of n. The mass matrix in the dual description is given by

$$\mathcal{M}_{ij} = M^2\begin{bmatrix} 1 & -q & 0 & \cdots & & & 0 \\ -q & 1+q^2 & -q & & & & \\ 0 & -q & 1+q^2 & \ddots & \ddots & & \vdots \\ & & \ddots & \ddots & \ddots & & \\ \vdots & & & \ddots & 1+q^2 & -q & 0 \\ & & & & -q & 1+q^2 & -q \\ 0 & & \cdots & & 0 & -q & r+q^2 \end{bmatrix}, \tag{4.16}$$

where $r = (\mu/M)^2$ is the square of the ratio between the two mass scales. The eigenmasses of this matrix are given by the roots of an $(N+1)$-th order polynomial. Again, we find that there is a tower of N massive modes whose masses go with the ultraviolet scale M. The remaining mode is massless in the limit where $r \to 0$ and is ultralight in general. We can find its mass by linearising the above eigenvalue problem; the resulting ultralight mass scale is given by

$$m_0^2 \simeq (q^2-1)^2 rM^2 \Big/ \big[q^{2(N+1)}(q^2+r-1) \\ + (N+1)r(1-q^2) - q^2 - r + 1\big]. \tag{4.17}$$

In the limit of large N (and for q larger than unity), m_0 is well approximated by

$$m_0^2 \simeq \frac{1}{q^{2(N+1)}}\frac{(q^2-1)^2 rM^2}{(q^2+r-1)}. \tag{4.18}$$

Thus, m_0 acquires an exponential suppression with respect to M, the argument in the exponential being the total number of clockwork gears, $N+1$. If we take $M = \mu = m_\text{P}$ ($m_\text{P} = 1.22 \times 10^{19}$ GeV being the Planck mass), $q = 2$ and $N = 200$, we get $m_0 \simeq 5.7 \times 10^{-33}$ eV, which is the energy scale associated with the Hubble rate. At

[1] Note that since the theory is quadratic in F this amounts to performing the Gaussian in the path integral, which can, of course, be done exactly.

large scales, the dynamics will be equivalent to quintessence driven by a quadratic potential. However, the mass scale of potential has not been tuned to the tiny value demanded by nature, rather it has arisen naturally on account of the clockwork mechanism and the coupling to the four-form. The underlying theory is made up uniquely of high scale couplings. We emphasize the fact that the mass mixings need not be identical but can have site dependence. As long as they are greater than unity, the clockwork mechanism will kick in as usual, even for order one couplings, and the suppression of the mass of the ultralight mode will occur as desired.

Finally we note that if we allowed for a non-vanishing flux, the potential would actually go as

$$V_0 = \frac{1}{2} m_0^2 \left(\frac{2\pi \mathcal{N}_Q e}{m_0} + \pi_0 \right)^2 . \tag{4.19}$$

For natural values of $\pi_0 \sim m_P$, the flux term dominates the potential for non-vanishing $\mathcal{N}_Q$ and there is too much dark energy. This is why we assumed a vanishing flux $\mathcal{N}_Q = 0$, which is a robust condition provided the nucleation rate for bubbles of non-vanishing flux is suppressed.[2]

4.4 Warm-Up: A String Motivated Toy Model

In this section and the next one, we would like to study the possibility of a UV embedding in string theory for the model Eq. (4.15). Although the question regarding the viability of a full embedding is complicated, we begin by considering the following: *can clockwork be embedded in fluxed type IIA perturbative supergravity?* The benefit of this setup is that, by turning all possible fluxes, one is allowed to fix all moduli at tree-level, with a number of complex structure moduli requiring non-geometric fluxes. The introduction of non-geometric fluxes does not come for free. Indeed, the fluxes backreact on the geometry leading to a half-flat non-Calabi Yau metric [29]. This creates a major caveat with regards to the trustability of these solutions. Nonetheless, we would like to move on and see if we can learn anything regarding a possible UV completion. Thus, in this section, we will see how the clockwork mechanism constraints the geometry of the internal space for a string motivated toy model. In the next section, we will study the scalar potential of fluxed type IIA and we will be able to provide a no-go for a supersymmetric type IIA clockwork, any embedding will require non-perturbative effects. Other considerations regarding the difficulty on clockwork embeddings can be found in [30].

---

[2] Although a detailed analysis of this interesting question is beyond the scope of our work, we expect that there is indeed suppression for the following reason. For $e \sim m_P^2$ a unit of flux generates a large positive Planckian potential. Therefore, we can crudely model the transition from vanishing to non-vanishing flux in terms of a bubble of Planckian de Sitter curvature nucleating in a quasi de Sitter background with a very small curvature. Such processes, tunnelling from true to false vacuum, are possible but are known to be suppressed relative to those going from false to true, becoming infinitely suppressed in the limit that the low scale curvature approaches zero (see e.g. [28]).

Our model can be motivated from the point of view of a higher dimensional theory. Indeed, in a theory with extra dimensions, a large number of scalar fields in the four-dimensional EFT is often associated with the periods of differential p-forms living in the higher dimensional theory. In [31, 32], the authors showed that one may rewrite the democratic type IIA supergravity formulation [33] in terms of a pseudo-action containing Minkowski 4-forms and its dual fields, which is equivalent to the democratic action at the level of the equations of motion. Further details on type IIA supergravity can be found in Appendix B.

The democratic action is usually written in terms of a polyform gauge invariant field strength, $G = G_0 + G_2 + \cdots + G_{10}$ where $\mathbf{G} = d\mathbf{C} + B \wedge \mathbf{C} + \mathcal{F} \wedge e^B$ and B is the Kalb-Ramond two-form, $\mathbf{C} = C_1 + C_3 + \cdots + C_9$ the polyform gauge field and $\mathcal{F} = \mathcal{F}_0 + \mathcal{F}_2 + \cdots + \mathcal{F}_{10}$ a formal sum of internal fluxes only. Let us take the following ansatz for the non-vanishing supergravity fields

$$\begin{aligned} B_2 = b^a(x)\omega_a, \quad C_3 = c_3^0(x), \quad C_5 = c_3^a(x)\omega_a\,, \\ C_7 = \tilde{c}_{3a}(x)\tilde{\omega}^a, \quad \mathcal{F}_2 = q^a\omega_a, \quad \mathcal{F}_4 = e_a\tilde{\omega}^a, \end{aligned} \tag{4.20}$$

where we note that this corresponds to the massless limit of type IIA supergravity. We have introduced the cohomology basis of {two, four}-forms in the internal manifold $\mathcal{M}_6$ as $\{\omega_a, \tilde{\omega}^a\}$, respectively, and ω_6 will denote the volume form of $\mathcal{M}_6$. With this ansatz, $\mathbf{G}$ reads[3]

$$\begin{aligned} G_2 = q^a\omega_a, \qquad G_4 = F_4^0 + \left(e_a + \mathcal{K}_{abc}b^bq^c\right)\tilde{\omega}^a, \\ G_6 = F_4^a\omega_a + \left(e_ab^a + \frac{1}{2}\mathcal{K}_{abc}b^ab^bq^c\right)\omega_6, \quad G_8 = \tilde{F}_{4a}\tilde{\omega}^a, \end{aligned} \tag{4.21}$$

with $\mathcal{K}_{abc} = \int_{\mathcal{M}_6}\omega_a \wedge \omega_b \wedge \omega_c$, the triple intersection numbers of $\mathcal{M}_6$.

Below we describe the effective four-dimensional action. In doing so, we will assume that the volume moduli, including the dilaton, are stabilised by some other ingredient of the theory (see e.g. [34]). Our main interest is to derive a plausible clockwork mechanism after compactifying the theory down to four dimensions.

These fields enter the four-dimensional potential as follows (see for details [31, 32])

$$\begin{aligned} V_{4D} \propto F_4^0 \wedge \star F_4^0 + g_{ab}F_4^a \wedge \star F_4^b + g^{ab}\tilde{F}_{4a} \wedge \star\tilde{F}_{4b} \\ +F_4^0\rho_0 + F_4^a\rho_a + \tilde{F}_{4a}\tilde{\rho}^a\,, \end{aligned} \tag{4.22}$$

where $g_{ab} = \int_{\mathcal{M}_6}\omega_a \wedge \star\omega_b$, $g^{ab} = \int_{\mathcal{M}_6}\tilde{\omega}^a \wedge \star\tilde{\omega}^b$ and we have introduced the dual scalars to the Minkowski 4-forms

---

[3] We have chosen to use the compact notation

$$F_4^0 = dc_3^0, \quad F_4^a = dc_3^a + b^aF_4^0, \quad \tilde{F}_{4a} = d\tilde{c}_{3a} + \mathcal{K}_{abc}b^bdc_3^c.$$

$$\rho_0 = e_a b^a + \frac{1}{2}\mathcal{K}_{abc} q^a b^b b^c ,$$
$$-\rho_a = e_a + \mathcal{K}_{abc} q^b b^c , \quad \tilde{\rho}^a = q^a .$$

Eliminating F_4^a and $\bar{F}_{4a}$ through their equations of motion in favour of their dual scalars, the four-dimensional Lagrangian density with only one Minkowski four-form is given by

$$\mathcal{L}_{4D} = -\frac{1}{\kappa_4^2}\Bigg[\frac{e^{-2\phi}}{4} g_{ab} db^a \wedge \star db^b + \frac{\mathcal{V}^3 e^{\frac{\phi}{2}}}{32} F_4^0 \wedge \star F_4^0$$
$$-\frac{1}{4} F_4^0 \rho_0 + \frac{e^{\frac{\phi}{2}}}{2\mathcal{V}^2}\left(g^{ab}\rho_a\rho_b + e^{\phi} g_{ab}\tilde{\rho}^a\tilde{\rho}^b\right)\Bigg] . \tag{4.23}$$

As this stands, the vacuum expectation value for the axions is not necessarily vanishing, owing to the presence of a tadpole in the potential for b_a. We can fix this by shifting $b_a \to \beta_a + b_a$ for some constant flux β_a chosen so that it eliminates the tadpole. By comparison to Eq. (4.14), one finds a Lagrangian density of the form[4]

$$\mathcal{L} = -\frac{1}{2}\left[\gamma_{ab}\partial b^a \partial b^b + M_{ab} b^a b^b + \frac{\alpha}{48}\left(F_4^0\right)^2\right] + \frac{1}{24}\chi_a b^a F_4^0 , \tag{4.24}$$

with

$$\gamma_{ab} = \frac{e^{-2\phi}}{2\kappa_4^2}\int_{\mathcal{X}} \omega_a \wedge \star\omega_b , \quad \alpha = \frac{2\mathcal{V}^3 e^{\frac{\phi}{2}}}{3\kappa_4^2} ,$$

$$M_{ab} = \frac{e^{\frac{\phi}{2}}}{\mathcal{V}^2\kappa_4^2}\mathcal{K}_{ade}\mathcal{K}_{bcf} q^d q^c \int_{\mathcal{X}} \tilde{\omega}^e \wedge \star\tilde{\omega}^f ,$$

$$\chi_a = \frac{6}{\kappa_4^2}\left(e_a + \mathcal{K}_{abc}\beta^b q^c\right) .$$

To realise the clockwork dark energy we need two ingredients: a mass matrix with vanishing eigenvalue and a zero mode that overlaps with the remaining four-form. The former is the clockwork condition while the latter allows the zero mode to acquire a small mass. If s^a is an eigenvector of M_{ab} with null eigenvalue, then the first condition requires $M_{ab}s^a \sim \mathcal{K}_{abc} q^c s^a = 0$ while the second gives $s^a \chi_a \neq 0$. To obtain the clockwork structure, we must therefore impose the following condition on the internal geometry

[4] In writing Eq. (4.24) we have dropped a next-to-leading order correction term to the coupling ($\sim F_4^0 b^2$) as well as a cosmological constant term and a total derivative that do not contribute to our discussion.

$$\int_{\mathcal{X}} \left(\omega_a s^a\right) \wedge \omega_b \wedge \left(\omega_c q^c\right) = 0\,. \tag{4.25}$$

A simple solution to the above constraint is provided by $s^a = q^a$ and the corresponding two-form given by a product of 1-forms, i.e. $\omega_c q^c = u \wedge v$. In that case, the constraint is immediately satisfied due to the antisymmetry of the wedge product on odd forms. As long as we require $q^a e_a \neq 0$, the second condition can also be satisfied.

The proposed solution could be heuristically realised if we consider our internal manifold to be a product $\mathcal{M}_6 \simeq T^2 \times \Sigma_g^4$, where g indicates the genus of the two-dimensional complex surface, which is assumed to be Kähler. Indeed, with this setup one naturally selects the zero mode, q^i, to point along the toroidal directions T^2. This is because the top form along T^2 is automatically a two-form corresponding to a product of the 1-forms of the torus. As long as we take the e_i to overlap slightly with the T^2, but not completely, we also satisfy the second condition, allowing the zero mode to gain a small mass through its four-form mixing. We have also shown that for the clockwork mechanism to bring the mass of the lightest mode down to the dark energy scale one requires $\mathcal{O}\,(200)$ scalar fields. For simplicity of exposition, let us think of these fields as descending only from the Kalb-Ramond two-form. Then, the previous requirement translates into a constraint on the cohomology structure of Σ_g^4 after fixing the T^2 geometry, namely $g = b_1 - b_2/2$ where b_p is the *p-th* Betti number of Σ_g^4. Using, $b_2 \sim \mathcal{O}\,(200)$ we find that $g \sim b_1 - 100$.

4.5 A More In-Depth Look into Achieving the Embedding

Although the string motivated scenario of the last section seems exciting, we can now show that the perturbative embedding is not viable. Indeed, consider a potential of the form [31]

$$V = \frac{1}{s^4}\left[\frac{\rho_0^2}{2}\mathcal{V} + \frac{g^{ab}\rho_a\rho_b}{8\mathcal{V}} + 2\mathcal{V} g_{ab}\tilde{\rho}^a\tilde{\rho}^b + \frac{\mathcal{V}}{2}\rho_m^2 - \frac{1}{6\mathcal{V}}\left(u^\lambda h_\lambda\right)^2\right] + \frac{h_0^2}{2\mathcal{V}}\frac{1}{s^2}\,. \tag{4.26}$$

This scalar potential is obtained from the usual dimensional reduction of the fluxed type IIA supergravity action[5] (further details are in Appendix B.4 and references therein). We have also allowed for a non-zero Romans mass term m in the type IIA action [35] such that the dual scalar are now defined as

[5] Up to some overall constant $R = 16e^{2K^{cs}}\mathrm{Re}(Z_0)^4$ that we have renormalised to unity since it will not affect the discussion of this section.

$$\rho_0 = \varphi + e_a b^a + \frac{1}{2}\mathcal{K}_{abc} q^a b^b b^c - \frac{m}{6}\mathcal{K}_{abc} b^a b^b b^c,$$
$$-\rho_a = e_a + \mathcal{K}_{abc} q^b b^c - \frac{m}{2}\mathcal{K}_{abc} b^b b^c, \quad \tilde{\rho}^a = q^a - m b^a , \quad \rho_m = m , \qquad (4.27)$$

let us also introduce the short hand notation $\hat{e}_a := e_a + \mathcal{K}_{abc} q^b b^c$ The scalar potential Eq. (4.26) contains s, the real part of the axio-dilaton, u^λ the $h^{1,1}$ complex structure saxions, φ a linear combination of axions, and b^a the $h^{2,1}$ Kähler axions. We will assume that $\mathcal{V}$ is a free parameter since the volume moduli can be fixed through fluxes to *a priori* large values (see Appendix B.4, and also [36]).

We concentrate on supersymmetric vacua where the Kähler F-terms are given in Eqs. (B.86) and (B.88) and fix

$$\tilde{\rho}^a = \rho_0 = 0 , \qquad 10\rho_a + 3m\mathcal{K}_{abc} v^b v^c = 0 , \qquad b^a := \frac{q^a}{2m} , \qquad (4.28)$$

Let

$$h_{ab} := -\frac{3m}{10}\mathcal{K}_{abc} v^c , \qquad (4.29)$$

so that

$$\rho_a = h_{ab} v^b , \quad \mathcal{V} = -\frac{5}{9m}\rho_a v^a . \qquad (4.30)$$

The kinetic metric for the Kähler sector can be written in terms of the Kähler potential in Eq. (2.63) as

$$g_{ab} = \frac{1}{4}\partial_{v^a}\partial_{v^b} K^{\text{ks}} = \frac{\mathcal{V}_a \mathcal{V}_b}{4\mathcal{V}^2} - \frac{\mathcal{V}_{ab}}{4\mathcal{V}} = \frac{9}{4}\frac{\rho_a \rho_b}{(\rho \cdot v)^2} - \frac{3}{2}\frac{h_{ab}}{\rho \cdot v} , \qquad (4.31)$$

here $\rho \cdot v := \rho_a v^a$. Its inverse is given by

$$g^{ab} = -\frac{2}{3}(\rho \cdot v) h^{ab} - 2 v^a v^b . \qquad (4.32)$$

The mass matrix for the b^a fields takes the form

$$M^a_b = \frac{2\mathcal{V}}{s^4}\left(\delta^a_b - \frac{g^{ac} g^{de}}{16 m \mathcal{V}^2}\mathcal{K}_{bcd}\rho_e\right) , \qquad (4.33)$$

with the clockwork conditions requiring the existence of an eigenvector n^a with vanishing eigenvalues, *i.e.*

$$M^a_b n^b \propto \left(1 - \frac{3}{5}\right) n^a + \frac{9}{5} v^a \frac{n \cdot \rho}{v \cdot \rho} = 0 . \qquad (4.34)$$

The only option to satisfy the eigenvalue condition is to choose $n^a = kv^a$ for some constant k, but then $M^a_b n^b = 0 \Rightarrow k = 0$ and there is no eigenvector satisfying this condition. Therefore, the clockwork conditions cannot be satisfied at the level of fluxed type IIA supergravity. It is then clear that, if clockwork is to be embedded in string theory, we will require the use of non-perturbative effects.

4.6 Deconstructing Clockwork Dark Energy

In the final section of this chapter, we provide a description of a braneworld construction of our clockwork dark energy model. If the number of pseudo-scalar fields is very high, $N \to \infty$, the clockwork mechanism can arise as a deconstruction of an extra compact dimension [14]. In this limit, the clockwork gears merge into a single field Φ, and the gear ϕ_i becomes the value of Φ at site i; the interaction of the four-form with the last site of the discrete clockwork corresponds to the localisation of the four-form on a brane at the boundary of the compactified extra dimension. Our model can be obtained from the following five-dimensional theory, defined on a fixed geometrical background,

$$S_{5D} = S_{\text{bulk}} + S_0 + S_{\pi R} \ , \tag{4.35}$$

where we have a canonical scalar in the bulk

$$S_{\text{bulk}} = \int d^4x \int_0^{\pi R} dy \, \sqrt{-g} \left[-\frac{m}{2} \, g^{IJ} (\partial_I \Phi)(\partial_J \Phi) \right] , \tag{4.36}$$

and two branes, one containing the dark energy sector,

$$S_0 = \int_{y=0} d^4x \sqrt{-\gamma_0} \left(\frac{\mu}{24} \Phi \frac{\epsilon^{\mu\nu\alpha\beta}}{\sqrt{-\gamma_0}} F_{\mu\nu\alpha\beta} - \frac{1}{48} F^{\mu\nu\alpha\beta} F_{\mu\nu\alpha\beta} \right) \Big] , \tag{4.37}$$

and the other containing the matter sector

$$S_{\pi R} = \int_{y=\pi R} d^4x \sqrt{-\gamma_R} \mathcal{L}_m (\gamma_R^{\mu\nu}, \Psi) \ . \tag{4.38}$$

We define g_{IJ} to be the five-dimensional metric, with indices I, J running over the four spacetime dimensions x^μ, as well as the additional fifth dimension y. For its part, the y dimension extends from 0 to πR, with a reflection symmetry at $y = 0$ and $y = \pi R$. These two boundary surfaces represent the location of the branes with the induced metric on the brane at $y = 0, \pi R$ given by $\gamma^0_{\mu\nu}$, $\gamma^R_{\mu\nu}$ respectively. Upon dividing the y dimension into $N + 1$ sites and discretising the action Eq. (4.35) accordingly, one can recover the Lagrangian density Eq. (4.14), provided we input the following geometry:

$$ds^2 = e^{\frac{4ky}{3}}(dy^2 + dx^2), \tag{4.39}$$

where dx^2 entails the four-dimensional metric on a brane, and identifying $q^N = e^{\pi k R}$ with the mass scale $M = N/(\pi R)$. We imagine all such scales - M, m, μ and k - to be of the same order, corresponding to the UV scale of the underlying braneworld theory. For this particular geometry, the latter corresponds to a five dimensional linear dilaton model with boundary terms living on the two branes [37].

However, the mechanism that suppresses the mass of the lightest mode is very general and works also for metrics that differ from Eq. (4.39). For instance, it works also in the following family of metrics:

$$ds^2 = e^{\frac{4ky}{3}}\left(e^{-4\ell k y}\, dy^2 + dx^2\right). \tag{4.40}$$

The clockwork geometry is recovered by choosing $\ell = 0$, while for $\ell = 1/3$ one gets Randall-Sundrum [38]. The equation of motion for Φ, once F has been integrated out, is

$$e^{2(\ell-\frac{1}{3})ky}\left[\Phi'' + 2(1+\ell)k\Phi' + \frac{\Box\Phi}{e^{4k\ell y}}\right] = \delta(y)\frac{\mu(\mu\Phi + Q)}{m}, \tag{4.41}$$

where a prime stands for a y derivative, and $\Box$ is the four-dimensional d'Alembertian. Setting $\Phi = -Q/\mu + \delta\Phi$, we can easily solve for $\delta\Phi$ in the bulk by taking the four-dimensional Fourier transform of Eq. (4.41)[6]. The solution is then given by

$$\begin{aligned}\delta\Phi(y, x^\mu) = \int d^4p\Big[A(p^2)\, J_{\frac{1+\ell}{2\ell}}\Big(\frac{\sqrt{-p^2}}{2k\ell e^{2k\ell y}}\Big) \\ + B(p^2)\, J_{-\frac{1+\ell}{2\ell}}\Big(\frac{\sqrt{-p^2}}{2k\ell e^{2k\ell y}}\Big)\Big]e^{ip_\mu x^\mu - k(1+\ell)y},\end{aligned} \tag{4.42}$$

where the $J_\alpha(z)$ are Bessel functions of the first kind, A and B being free functions. Equation Eq. (4.41) further imposes some boundary conditions at $y = 0$ and πR. As a result, A and B are linearly related, while p^2 can only take some values among a quantised set, $p^2 = -m_n^2$, $n \in \mathbb{N}$. The masses m_n are found as solutions of the following equation:

---

[6] In general, one should embed the action Eq. (4.35) in some geometric theory and study the perturbations of the geometric quantities as well as the fluctuations of the clockwork field. However, these fluctuations decouple at linear level. We can thus consistently examine the fluctuations of Φ on their own.

$$
\begin{aligned}
&J_{\frac{\ell-1}{2\ell}}\left(\frac{m_n}{2k\ell}e^{-2k\ell\pi R}\right)\left\{J_{\frac{\ell+1}{2\ell}}\left(\frac{m_n}{2k\ell}\right)\left[4k(\ell+1)m+\mu^2\right]\right.\\
&\left.-2m_n m J_{\frac{3\ell+1}{2\ell}}\left(\frac{m_n}{2k\ell}\right)\right\}+J_{\frac{1-\ell}{2\ell}}\left(\frac{m_n}{2k\ell}e^{-2k\ell\pi R}\right)\\
&\left[\mu^2 J_{-\frac{\ell+1}{2\ell}}\left(\frac{m_n}{2k\ell}\right)-2m_n m J_{\frac{\ell-1}{2\ell}}\left(\frac{m_n}{2k\ell}\right)\right]=0\,.
\end{aligned}
\tag{4.43}
$$

They are all of order k, except for the first which is very light, being suppressed by $e^{k\pi R} \gg 1$. It is possible to find an approximate expression for this light mass:

$$
\begin{aligned}
m_0^2(\ell) \simeq &\frac{\mu^2}{m} 8k^2\ell(1-\ell^2) \Big/ \left\{2e^{2k(1-\ell)\pi R}\ell\left[2k(1+\ell)+\frac{\mu^2}{m}\right]\right.\\
&\left.-(1+\ell)\left(4k\ell+\frac{\mu^2}{m}\right)-e^{-4k\ell\pi R}(\ell-1)\frac{\mu^2}{m}\right\}.
\end{aligned}
\tag{4.44}
$$

The above approximation is based on an expansion of the Bessel functions assuming $m_0 \ll k\ell$. This is no longer true when $\ell \to 0$. To compute the lightest mass in the clockwork geometry, one can solve directly

$$
m_0^2(0) = \frac{4k^2\mu^2}{e^{2\pi kR}\left(4km+\mu^2\right)-4km-2\pi k\mu^2 R+\mu^2}\,.
$$

Equation Eq. (4.44) shows that the suppression mechanism works for the whole family of metrics Eq. (4.40), even if it gets less efficient as we depart from the clockwork geometry. Note that Eq. (4.44) should not be trusted physically when $l \gtrsim 1$, because the length of the y dimension drops below the UV scale of the five dimensional theory.

As it stands, the mass of the lightest dark energy mode is suppressed relative to UV scales set by matter resident on *either* of the two branes. Why, then, have we placed matter on the brane at $y = \pi R$? This is because the energy density of the dark energy field in slow roll is enhanced by the effective four dimensional Planck scale. This enhancement exactly compensates for the suppression in the mass of the field, and in the end the energy density during slow roll scales like k^4. In any event, it turns out that the energy density of dark energy will only be suppressed if we calibrate our scales relative to the πR brane, where the warp factor is exponentially large. This is why we put the visible sector of our theory on the right hand brane.

4.7 Conclusions

In this chapter, we have studied the possibility of generating naturally small masses from Planckian scales. In particular, we have considered how a parametrically small dark energy field could arise in a (technically) natural way. By considering the interplay between clockwork gravity and the coupling between its pseudo-scalar fields and Minkowski top-forms we were able to generate a super-light dark energy field that arises solely from the characteristics of the high energy theory.

Furthermore, we have explored how this mechanism could be achieved from the point of view of a prospective UV completion of the theory. We have studied a type IIA toy model and translated the existence of a vanishing eigenvalue in the clockwork mass matrix as a constraint in the internal geometry. Even more so, we have been able to show that, after considering the fluxed perturbative type IIA scalar potential, a no-go arises. Our hopes of generating a stringy embedding of clockwork will then have to pass through the introduction of corrections to the tree-level action. Since non-perturbative effects are better understood in the type IIB dual [39], we will further only concern ourselves with type IIB supergravity in the remainder of the thesis. Finally, we were also able to write down a generalised braneworld scenario, motivated by the work of [14], where the DKS mode lives on a brane.

References

1. Bordin L, Cunillera F, Lehébel A, Padilla A (2020) A natural theory of dark energy. Phys Rev D 101:085012. https://doi.org/10.1103/PhysRevD.101.085012, arxiv: 1912.04905
2. Choi K, Im SH (2016) Realizing the relaxion from multiple axions and its UV completion with high scale supersymmetry. JHEP 14901:149. https://doi.org/10.1007/JHEP01, arxiv: 1511.00132
3. Kaplan DE, Rattazzi R (2016) Large field excursions and approximate discrete symmetries from a clockwork axion. Phys Rev D93:085007. https://doi.org/10.1103/PhysRevD.93.085007, arxiv: abs/1511.01827
4. Dvali G (2006) Large hierarchies from attractor vacua. Phys Rev D 74:025018. https://doi.org/10.1103/PhysRevD.74.025018. arxiv: abs/hep-th/0410286hep-th/0410286
5. Dvali G Three-form gauging of axion symmetries and gravity. arxiv: abs/hep-th/0507215hep-th/0507215
6. Kaloper N, Sorbo L (2009) A natural framework for Chaotic Inflation. Phys Rev Lett 102:121301. https://doi.org/10.1103/PhysRevLett.102.121301, arxiv: 0811.19890
7. Kaloper N, Lawrence A, Sorbo L (2011) An ignoble approach to large field inflation. JCAP 03:023. https://doi.org/10.1088/1475-7516/2011/03/023, arxiv: 1101.00261
8. Banks T, Dine M, Fox PJ, Gorbatov E (2003) On the possibility of large axion decay constants. JCAP06:001. https://doi.org/10.1088/1475-7516/2003/06/001, arxiv: abs/hep-th/0303252hep-th/0303252
9. Kim JE, Nilles HP, Peloso M (2005) Completing natural inflation. JCAP 01:005. https://doi.org/10.1088/1475-7516/2005/01/005, arxiv: abs/hep-ph/0409138hep-ph/0409138
10. Arkani-Hamed N, Motl L, Nicolis A, Vafa C (2007) The String landscape, black holes and gravity as the weakest force. JHEP 06:060. https://doi.org/10.1088/1126-6708/2007/06/060, arxiv: abs/hep-th/0601001hep-th/0601001
11. Brown J, Cottrell W, Shiu G, Soler P (2015) Fencing in the swampland: quantum gravity constraints on large field inflation. JHEP 10:023. https://doi.org/10.1007/JHEP10(2015)023, arxiv: abs/1503.04783
12. Cheung C, Remmen GN (2014) Naturalness and the weak gravity conjecture. Phys Rev Lett 113:051601. https://doi.org/10.1103/PhysRevLett.113.051601, arxiv: abs/1402.22871
13. Rudelius T (2015) Constraints on axion inflation from the weak gravity conjecture. JCAP 09:020. https://doi.org/10.1088/1475-7516/2015/9/020, arxiv: abs/1503.00795

14. Giudice GF, McCullough M (2017) A clockwork theory. JHEP 02:036. https://doi.org/10.1007/JHEP02(2017)036, arxiv: abs/1610.07962
15. ’t Hooft G (1980) Naturalness, chiral symmetry, and spontaneous chiral symmetry breaking. NATOSci Ser B 59:135. https://doi.org/10.1007/978-1-4684-7571-5_9
16. Abbott LF (1985) A mechanism for reducing the value of the cosmological constant. Phys Lett B 150:427. https://doi.org/10.1016/0370-2693(85)90459-9
17. Brown JD, Teitelboim C (1985) Dynamical neutralization of the cosmological constant. Phys Lett B 195:177. https://doi.org/10.1016/0370-2693(87)91190-7
18. Brown JD, Teitelboim C (1988) Neutralization of the cosmological constant by membrane creation. Nucl Phys B 297:787. https://doi.org/10.1016/0550-3213(88)90559-7
19. Silverstein E, Westphal A (2008) Monodromy in the CMB: gravity waves and string inflation. Phys Rev D 78:106003. https://doi.org/10.1103/PhysRevD.78.106003, arxiv: abs/0803.30850
20. McAllister L, Silverstein E, Westphal A (2010) Gravity waves and linear inflation from axion monodromy. Phys Rev D 82:046003. https://doi.org/10.1103/PhysRevD.82.046003, arxiv: abs/0808.07060
21. Valenzuela I (2017) Backreaction issues in axion monodromy and Minkowski 4-forms. JHEP06:098. https://doi.org/10.1007/JHEP06(2017)098, arxiv: /abs/1611.003941611.00394
22. Kaloper N, Lawrence A (2017) London equation for monodromy inflation. Phys Rev D95:063526. https://doi.org/10.1103/PhysRevD.95.063526, arxiv: abs/1607.06105
23. Padilla A (2019) Monodromy inflation and an emergent mechanism for stabilising the cosmological constant. JHEP 01:175. https://doi.org/10.1007/JHEP01(2019)175, arxiv: abs/1806.04740
24. El-Menoufi BK, Nagy S, Niedermann F, Padilla A (2019) Quantum corrections to vacuum energy sequestering (with monodromy). Class Quant Grav 36:215014. https://doi.org/10.1088/1361-6382/ab46f6, arxiv: abs/1903.07612
25. Kaloper N (2019) Irrational monodromies of vacuum energy. JHEP 11:106. https://doi.org/10.1007/JHEP11(2019)106, arxiv: abs/1806.03308
26. Polchinski J, Strominger A (1996) New vacua for type II string theory. Phys Lett B 388:736. https://doi.org/10.1016/S0370-2693(96)01219-1, arxiv: abs/hep-th/9510227hep-th/9510227
27. Bousso R, Polchinski J (2000) Quantization of four form fluxes and dynamical neutralization of the cosmological constant. JHEP 06:006. https://doi.org/10.1088/1126-6708/2000/06/006, arxiv: abs/hep-th/0004134hep-th/0004134
28. Lee K-M, Weinberg EJ (1987) Decay of the true vacuum in curved space-time. Phys Rev D36:1088. https://doi.org/10.1103/PhysRevD.36.1088
29. Acharya BS, Benini F, Valandro R (2007) Fixing moduli in exact type IIA flux vacua. JHEP 02:018. https://doi.org/10.1088/1126-6708/2007/02/018, arxiv: abs/hep-th/0607223hep-th/0607223
30. Ibanez LE, Montero M (2018) A note on the WGC, effective field theory and clockwork within string theory. JHEP 02:057. https://doi.org/10.1007/JHEP02(2018)057, arxiv: abs/1709.02392
31. Bielleman S, Ibanez LE, Valenzuela I (2015) Minkowski 3-forms, flux string vacua, axion stability and naturalness. JHEP 12:119. https://doi.org/10.1007/JHEP12(2015)119, arxiv: abs/1507.06793
32. Carta F, Marchesano F, Staessens W, Zoccarato G (2016) Open string multi-branched and Kähler potentials. JHEP 09:062. https://doi.org/10.1007/JHEP09(2016)062, arxiv: abs/1606.00508
33. Bergshoeff E, Kallosh R, Ortin T, Roest D, Van Proeyen A (2001) New formulations of D = 10 supersymmetry and D8 - O8 domain walls. Class Quant Grav 18:3359. https://doi.org/10.1088/0264-9381/18/17/303, arxiv: abs/hep-th/0103233hep-th/0103233
34. Saltman A, Silverstein E (2006) A new handle on de Sitter compactifications. JHEP 01:139. https://doi.org/10.1088/1126-6708/2006/01/139, arxiv: abs/hep-th/0411271hep-th/0411271
35. Romans LJ (1986) Massive N=2a Supergravity in Ten-Dimensions. Phys Lett B 169:374. https://doi.org/10.1016/0370-2693(86)90375-8
36. DeWolfe O, Giryavets A, Kachru S, Taylor W, Type IIA moduli stabilization. JHEP 07:066. https://doi.org/10.1088/1126-6708/2005/07/066, arxiv: abs/hep-th/0505160hep-th/0505160

37. Giudice GF, McCullough M, Comment on "disassembling the clockwork mechanism". arxiv: abs/1705.10162
38. Randall L, Sundrum R (1999) A large mass hierarchy from a small extra dimension. Phys Rev Lett 83:3370. https://doi.org/10.1103/PhysRevLett.83.3370, arxiv: abs/hep-ph/9905221hep-ph/9905221
39. Cicoli M, De Alwis S, Maharana A, Muia F, Quevedo F (2019) De Sitter vs Quintessence in string theory. Fortsch Phys 67:1800079. https://doi.org/10.1002/prop.201800079, arxiv: abs/1808.08967

Chapter 5
A Stringy Perspective on the Coincidence Problem

In the last chapter, we tried to tackle the idea of how to obtain naturally small couplings in a theory with only $\mathcal{O}(M_{\rm pl})$ parameters. In this one, we turn our attention to the coincidence problem discussed in Chap. 3. In [1], we argued that, for string compactifications broadly consistent with swampland constraints, dark energy is likely to signal the beginning of the end of our universe as we know it, perhaps even through decompactification, with possible implications for the cosmological coincidence problem. Thanks to the scarcity (absence?) of stable de Sitter vacua, dark energy in string theory is assumed to take the form of a quintessence field in slow roll. As it rolls, a tower of heavy states will generically descend, triggering an apocalyptic phase transition in the low energy cosmological dynamics after at most a few hundred Hubble times. As a result, dark energy domination cannot continue indefinitely and there is at least a percentage chance that we find ourselves in the first Hubble epoch. We begin by describing the some ideas borrowed from Sect. 2.5 that make the core of our discussion and, in Sect. 5.2, we use a toy model of quintessence coupled to a tower of heavy states to explicitly demonstrate the breakdown in the cosmological dynamics as the tower becomes light.

5.1 Generic Idea

We begin with the refined Swampland distance conjecture (rSDC) [2–5], which is one of the most well studied and least controversial of the swampland conjectures (see also [6–13]). As we described in Chap. 2, the rSDC states the following: consider two points in field space, ϕ_0 and $\phi_0 + \Delta\phi$, separated by a geodesic distance $\Delta\phi$. As $\Delta\phi \to \infty$, there exists an infinite tower of states whose mass become exponentially light,

$$m(\phi_0 + \Delta\phi) \sim m(\phi_0) e^{-\beta \frac{|\Delta\phi|}{M_{Pl}}} \tag{5.1}$$

F. Cunillera, *Dark Energy*, Springer Theses,
https://doi.org/10.1007/978-3-031-21468-4_5

for some positive constant, β, that we typically expect to be $\mathcal{O}(1)$. The offending tower of states is often associated with Kaluza-Klein modes or winding modes, depending on the direction of motion in moduli space —although in some cases the tower can originate from localised sources that exist in the theory, see for example [14]. For this reason, we take the initial mass $m(\phi_0)$ to be given by the scale of compactification, $1/R$, which could be as low as a few meV in a braneworld setting where Standard Model fields are confined to a 3-brane, although generically we expect it to be much larger, perhaps even just short of the Planck scale, $M_{Pl} \sim 10^{18}$ GeV.

In its refined form, the de Sitter swampland conjecture (rdSC) [15–17] concerns the form of the potential $V(\phi)$ for scalar fields in a low energy effective theory. Assuming the effective theory is obtained from a consistent theory of quantum gravity, the potential must satisfy either

$$|\nabla V| \geq \frac{c}{M_{Pl}} V \tag{5.2}$$

or

$$\min\left(\nabla_i \nabla_j V\right) \leq -\frac{c'}{M_{Pl}^2} V \tag{5.3}$$

where c, c' are universal positive constants of $\mathcal{O}(1)$ and $\min\left(\nabla_i \nabla_j V\right)$ is the minimum eigenvalue of the corresponding Hessian. Note that the conjecture forbids the existence of stable de Sitter vacua in string theory, for which we would have to have $V > 0$. It does, however, allow for de Sitter vacua with a tachyonic instability of order the corresponding Hubble time $H^{-1} \sim M_{Pl}/\sqrt{V}$. Some constraints on the scale of the tachyon were derived in the context of 10D supergravity [18].

We remark that we will not make use of the rdSC *per se*. Motivated by the difficulty to achieve de Sitter vacua in string theory, of which the rdSC is a symptom, we would like to ask the question of what does a dynamical model of dark energy look like for a generic string compactification. By generic we mean those string compactifications to which the rSDC applies. This is generic in so far that the rSDC applies to models in the *Dine-Seiberg region*, as discussed in Chap. 2. This leaves out models that require perturbative and non-perturbative corrections to the tree-level action, those living in the *pheno region* depicted in Fig. 2.2. An analysis of those models will be conducted in Chap. 8.

Of course, building a reliable model of quintessence within string theory is not without its own challenges [19, 20], and we will also explore these in detail in Chap. 8. Nevertheless, we begin with a model of quintessence as a canonical scalar field, ϕ, moving under the influence of a potential V. Here we imagine that quintessence is generically described by the saxions of string theory with a non-compact field range. The energy density and pressure stored in the field are given respectively by $\rho_\phi = \frac{1}{2}\dot{\phi}^2 + V$ and $p_\phi = \frac{1}{2}\dot{\phi}^2 - V$. The dynamics of the scalar is governed by the following field equation

$$\ddot{\phi} + 3H\dot{\phi} + V'(\phi) = 0 \tag{5.4}$$

where $H(t)$ is the Hubble parameter and "dot" denotes differentiation with respect to cosmological time. Our goal is to argue that dark energy domination is relatively short lived on account of the motion of the field towards the infinite points in moduli space. Any motion of the field prior to dark energy domination will only reduce the amount it is allowed to move afterwards, bringing dark energy to an even quicker conclusion. Therefore, the most conservative scenario is to assume negligible motion of the moduli fields until dark energy finally begins to dominate. This is, in any event, likely as the field will be held up by Hubble friction.

Once the dark energy field has come to dominate, to be compatible with the observed equation of state, it must be in slow roll, $\frac{1}{2}\dot{\phi}^2 \ll V$, $|\ddot{\phi}| \ll 3H|\dot{\phi}|$. Furthermore, we may assume that $H \approx H_0 \sim 10^{-33}$ eV $\sim 10^{-60} M_{Pl}$. With these approximations, consider the field excursion in a short time δt. This is given by

$$\delta\phi \approx -\frac{V'}{3H_0}\delta t \tag{5.5}$$

If we accept the refined de Sitter conjectures, then one of Eqs. (5.2) or (5.3) must hold. Let's assume we satisfy the condition on the gradient, given by Eq. (5.2). It now follows that

$$|\delta\phi| \gtrsim \frac{c}{M_{Pl}}\frac{V}{3H_0}\delta t \sim \mathcal{O}(1) M_{Pl} H_0 \delta t \tag{5.6}$$

where we have used the Friedmann equation during dark energy domination $H_0^2 \approx \frac{\rho_\phi}{3M_{Pl}^2} \approx \frac{V}{3M_{Pl}^2}$ and the fact that c is assumed to be $\mathcal{O}(1)$. We immediately see that the dark energy field rolls roughly a Planck unit in a Hubble time. We now consider the implications for the distance conjecture Eq. (5.1), assuming the tower of new states are initially very heavy, with masses $m(\phi_0) \sim M_{Pl}$ close to the Planck scale. After a single Hubble time, the dark energy field will move by around one Planck unit. This only corresponds to a fractional change in the masses in the tower, and certainly not enough to contaminate the low energy physics.

How far is the field allowed to roll before we have to start worrying about it? In the local neighbourhood of the Earth the field should not be displaced from ϕ_0 by more than around 30 Planck units. Anything more than that would bring the mass of the tower down from Planck scale to the scale of collider physics, opening up the possibility of producing these states at the LHC. Of course, the details of this depend on the nature of the coupling between the tower and the Standard Model fields. Furthermore, none of these considerations are relevant on cosmological scales, where we can certainly allow the field to move much further. To ensure that the tower remains decoupled from the low scale cosmological dynamics, we conservatively impose a maximum displacement of $|\Delta\phi|_{\max} \sim \frac{M_{Pl}}{\beta}\ln(M_{Pl}/H_0)$. For $\beta \sim \mathcal{O}(1)$ this corresponds to a displacement of around 140 Planck units. Assuming the field continues to roll a Planck unit in every Hubble time, we see that the tower of states will trigger a transition in the cosmological dynamics after no more than $\mathcal{O}(100)$ Hubble times. The coincidence problem isn't solved but it is significantly ameliorated. If a generic model of dark energy is destined to last at most $\mathcal{O}(100)$ Hubble epochs, and all epochs are equally probable, we might expect there to be at least a percentage chance that we find ourselves in the first epoch.[1]

What if we assume that we satisfy the second of the two de Sitter swampland criteria in Eq. (5.3), rather than Eq. (5.2)? If the condition on the gradient of Eq. (5.2) is violated, but the condition on the Hessian in Eq. (5.3) holds the dark energy field may move considerably less than a Planck unit in a Hubble time. In fact, if we imagined the field sitting in a region where the gradient of the potential is negligible, we might even imagine it staying there indefinitely, giving a neverending period of dark energy. However, this conclusion is too quick. Quantum fluctuations will guarantee a displacement in the field of at least $\mathcal{O}(H_0)$ in the first Hubble time. This initial displacement will grow thanks to the fact that the Hessian condition implies a tachyonic mass for the fluctuations in the dark energy field, $\mu^2 = V''(\phi) < 0$ with $|\mu^2| \gtrsim \frac{c'}{M_{Pl}^2} V \approx 3c' H_0^2$. For $c' \sim \mathcal{O}(1)$, the corresponding instability can be as slow as a Hubble time but even so, its effect is amplified over several Hubble times by exponential growth. There are two possibilities: the first is that the instability triggers a rapid transition which brings the acceleration to a premature end (as desired for the coincidence problem). This could occur, for example, by the potential changing sign so that we no longer have a quasi de Sitter expansion. Alternatively, the background cosmology could remain roughly unchanged, at least beyond a few Hubble times. If this is the case, the tachyonic instability amplifies the initial displacement to a value[2]

$$|\delta\phi| \gtrsim H_0 e^{\mathcal{O}(1) H_0 \delta t} \tag{5.7}$$

In the latter scenario, the tower of massive states would remain decoupled from the cosmological dynamics until $|\delta\phi| \approx |\Delta\phi|_{\max} \sim \frac{M_{Pl}}{\beta} \ln(M_{Pl}/H_0)$, at which point a transition is inevitable. For $\beta \sim \mathcal{O}(1)$, this will occur within at most $\ln[M_{Pl}/H_0 \ln(M_{Pl}/H_0))] \sim 143$ Hubble times, so our conclusions are unchanged, and the coincidence problem isn't as serious as we previously thought.

Of course, the de Sitter conjecture is less well established than the distance conjecture, and, in the simplest scenarios, may even be at odds with local measurements of H_0 [21]. With this in mind, what can we say if we deny the validity of both criteria Eqs. (5.2) and (5.3) and abandon the de Sitter conjecture altogether? In this instance, we cannot rule out the possibility that the current phase of acceleration is approaching a stable de Sitter configuration in which dark energy continues for an exponentially large number of Hubble epochs. If this is the case, then the coincidence

---

[1] By way of comparison, we note that when Leicester City won the premier league in 2016, they started the season as 5000-1 outsiders.

[2] To leading order, the scalar satisfies an equation $\ddot{\phi} + 3H_0\dot{\phi} - |\mu^2|\phi = 0$, where we recall that $\mu^2 = V''(\phi) < 0$ with $|\mu^2| \gtrsim \frac{c'}{M_{Pl}^2} V \approx 3c' H_0^2$. The general solution is then given by a sum of exponentials $e^{\lambda_\pm t}$ where $\lambda_\pm = \frac{-3H_0 \pm \sqrt{9H_0^2 + 4|\mu^2|}}{2}$, which includes a growing mode, with $\lambda_+ \gtrsim H_0$, thanks to the tachyonic instability with $|\mu^2| \gtrsim H_0^2$.

problem is as problematic as ever. However, we might tentatively speculate that the de Sitter conjecture is really a statement about what is *generic* within consistent models of dark energy within string theory. Of course, it is much too early to make any definitive statement in this regard. Nevertheless, if it happens to be true that the generic scenarios are those for which one of the criteria Eqs. (5.2) or (5.3) hold, our results go through and the coincidence problem is tamed. We might also worry about the fact we have assumed dark energy to be a single canonical scalar. However, we expect this to capture the generic dynamics of fields moving through moduli space, with our canonical scalar tangential to the trajectory and all the orthogonal directions stabilised.

5.2 A Toy Model

To better understand how an accumulation of light states can impact the cosmological evolution at late times, we consider a toy model of dark energy described by the following Lagrangian

$$\mathcal{L} = -\frac{1}{2}(\partial\phi)^2 - \mu^4 e^{-\frac{\alpha\phi}{M_{Pl}}} + \sum_{n=1}^{\infty} -\frac{1}{2}(\partial\varphi_n)^2 - \frac{1}{2}n^2 M_{KK}^2 e^{-2\frac{\beta\phi}{M_{Pl}}}\varphi_n^2 \tag{5.8}$$

Here ϕ is the quintessence field, driving dark energy, taken to be in slow roll on a runaway potential of the form $\mu^4 e^{-\frac{\alpha\phi}{M_{Pl}}}$ where $\mu^4 = 3M_{pl}^2 H_0^2$ is the scale of dark energy and α is some order one positive number. To be compatible with observations we require[3] $\alpha \lesssim 1.02$ [23], which is not in conflict with the swampland constraints on the potential Eqs. (5.2) and (5.3). In addition we have a tower of heavy states, φ_n, whose masses are originally set by some high scale M_{KK} (imagined to be the Kaluza-Klein scale associated with the compact internal manifold), becoming exponentially light as the dark energy field, ϕ, rolls off to infinity. The rate at which the tower becomes light is set by the coupling β, which is also assumed to be order one and positive.

At the dawn of dark energy domination, at time t_0, we assume that the quintessence field is far away from the tails of the exponentials, $\phi \sim \phi_0 \ll M_{Pl}/\alpha, M_{Pl}/\beta$. Since the effective mass of the Kaluza-Klein tower, $M_{KK}e^{-\frac{\beta\phi_0}{M_{Pl}}} \sim M_{KK} \gg H_0$ is high in this regime, far above the scale of the cosmological evolution, the Kaluza-Klein states are decoupled from the dynamics. As a result, the quintessence field satisfies the following classical equation of motion on a homogenous and isotropic background

[3] We should note that in [22] the authors claim a more stringent bound of $\alpha \lesssim 0.6$ at 3σ level. However, the more detailed analysis in [23] gives the more relaxed bound $\alpha \lesssim 1.02$ at 3σ, which is the one that we use in this section. For a detailed analysis of the two approaches we point the interested reader to [23].

$$\ddot{\phi} + 3H\dot{\phi} - \frac{\alpha}{M_{Pl}}\mu^4 e^{-\frac{\alpha\phi}{M_{Pl}}} = 0 \tag{5.9}$$

For a quasi-de Sitter expansion, with $H \approx H_0$, we find that $\phi \approx \phi_0 + \lambda(t - t_0)$ where

$$\lambda = \frac{\alpha\mu^4}{3M_{Pl}H_0} = \alpha M_{Pl}H_0 < M_{Pl}H_0 \tag{5.10}$$

This approximation works well as long as $\Delta\phi < M_{Pl}/\alpha$, or equivalently, $\Delta t < H_0^{-1}/\alpha^2$. Of course, acceleration will continue beyond this time, only at a lower scale. There is a wealth of literature on the dynamics of similar quintessence models (for a review, see [24]). In this chapter, we wish to briefly explore another effect that is far less well studied - the time dependence on the mass of the Kaluza-Klein tower as the field begins to slowly roll. On the quasi-de Sitter background with constant curvature, this is given by

$$M_{KK}^{\text{eff}}(t) = M_{KK}e^{-\frac{\beta s}{M_{Pl}}} \approx M_{KK}e^{-\epsilon(t-t_0)} \tag{5.11}$$

where $\epsilon = \beta\lambda/M_{Pl} = \alpha\beta H_0$. This will drive particle production in the Kaluza-Klein sector, kicking in as soon as the states stop being decoupled, $M_{KK}^{\text{eff}}(t) \lesssim H_0$. A complete analysis of this phenomena requires a detailed numerical study of particle creation on the dynamical background, taking into account the effect of the time varying mass and the de Sitter cosmology.

To get some immediate insight into what might happen we neglect the curvature of the background spacetime and focus on the particle production due to a mass varying exponentially with time on a Minkowski geometry. Crucially, these approximations allow us to make *preliminary* analytic estimates but we should also acknowledge their limitations. In particular, we are implicitly assuming that the dark energy field, which feeds into the effective mass of the Kaluza Klein tower, continues to evolve linearly in time beyond the first Hubble epoch. In truth, the Hubble scale changes and the dark energy field picks up additional temporal dependence which may affect some of the details. Also, we are neglecting the effect of spacetime curvature. This is less of an issue, as our interest here is on the effect of the changing mass of the Kaluza-Klein tower, as opposed to the effect of quantum fields propagating on de Sitter. We also expect this approximation to accurately capture the physics on sub-horizon scales.

With these caveats in mind, let us take a closer look at the nth state in the Kaluza-Klein tower on a Minkowski background, whose dynamics described by the following Lagrangian

$$\mathcal{L} = -\frac{1}{2}(\partial\varphi_n)^2 - \frac{1}{2}n^2 M_{KK}^2 e^{-2\epsilon t}\varphi_n^2 \tag{5.12}$$

where we have also set $t_0 = 0$ (without loss of generality). Our analysis follows the standard techniques reviewed in detail in [25], whose conventions we also follow. As explained in [25], the state operator for the quantum field can be expanded in terms of creation and annihilation operators, $\hat{a}_{\mathbf{k}}^\dagger$ and $\hat{a}_{\mathbf{k}}$, in the usual way,

$$\hat{\varphi}_n(t,\mathbf{x}) = \frac{1}{\sqrt{2}}\int \frac{d^3\mathbf{k}}{(2\pi)^{\frac{3}{2}}}\left[e^{i\mathbf{k}\cdot\mathbf{x}}\bar{u}_k(t)\hat{a}_{\mathbf{k}} + e^{-i\mathbf{k}\cdot\mathbf{x}}u_k(t)\hat{a}^\dagger_{\mathbf{k}}\right] \tag{5.13}$$

where "bar" denote the complex conjugate. The mode functions are governed by the equation for a time dependent harmonic oscillator

$$\ddot{u}_k + \omega_k^2(t)u_k = 0, \qquad \omega_k^2(t) = k^2 + n^2[M^{\text{eff}}_{KK}(t)]^2 \tag{5.14}$$

and have solutions that are conveniently expressed in terms of Hankel functions (of the first kind)

$$u_k(t) = AH^{(1)}_{i\mu}(z) + B\bar{H}^{(1)}_{i\mu}(z) \tag{5.15}$$

where $z = \frac{nM_{KK}}{\epsilon}e^{-\epsilon t}$ and $\mu = \frac{k}{\epsilon}$. Since $[M^{\text{eff}}_{KK}(t)]^2 > 0$, at any given time t_* we can define the instantaneous vacuum state $|0\rangle_*$ as the lowest energy state of the Hamiltonian at that time. The mode functions that determine this state satisfy the boundary condition [25]

$$u_k(t_*) = \frac{1}{\sqrt{\omega_k(t_*)}}, \qquad \dot{u}_k(t_*) = i\sqrt{\omega_k(t_*)} \tag{5.16}$$

This fixes the constants in Eq. (5.15) so that

$$A = \frac{\pi i e^{-\pi\mu}}{4\sqrt{\omega_k(t_*)}}\left[z_*\bar{H}^{(1)\prime}_{i\mu}(z_*) + \frac{\omega_k(t_*)}{\epsilon}i\bar{H}^{(1)}_{i\mu}(z_*)\right] \tag{5.17}$$

$$B = -\frac{\pi i e^{-\pi\mu}}{4\sqrt{\omega_k(t_*)}}\left[z_*H^{(1)\prime}_{i\mu}(z_*) + \frac{\omega_k(t_*)}{\epsilon}iH^{(1)}_{i\mu}(z_*)\right] \tag{5.18}$$

The mode functions that determine the "in" vacuum at the beginning of the dark energy era, $u^{\text{in}}_k(t)$, are given by these expressions for $A,\ B$ with the choice $t_* = 0$. The mode functions $u^{\text{out}}_k(t)$, that determine the "out" vacuum at some later time, T, are given by the same formulae but with $t_* = T$. As usual, the two can be related by a Bogoliubov transformation, of the form $u^{\text{in}}_k(t) = \alpha_k(T)u^{\text{out}}_k(t) + \beta_k(T)\bar{u}^{\text{out}}_k(t)$. At time $T > 0$, the true vacuum state differs from the initial vacuum state, and so the latter contains particles. As explained in [25], the mean particle number density for modes of momentum $\mathbf{k}$ is $N_k(T) = |\beta_k(T)|^2$ with corresponding energy density $E_k(T) = \omega_k(T)N_k(T)$. When we calculate this explicitly, it turns out that

$$N_k(T) = \frac{\pi^2}{16}e^{-2\pi\mu}\epsilon^2\left[X^2 + Y^2\right] \tag{5.19}$$

where

$$X = \text{Im}\left[\frac{z_{\text{in}}H^{(1)\prime}_{i\mu}(z_{\text{in}})}{\sqrt{\omega_k(0)}}\frac{z_{\text{out}}\bar{H}^{(1)\prime}_{i\mu}(z_{\text{out}})}{\sqrt{\omega_k(T)}} - \frac{\sqrt{\omega_k(0)}H^{(1)}_{i\mu}(z_{\text{in}})}{\epsilon}\frac{\sqrt{\omega_k(T)}\bar{H}^{(1)}_{i\mu}(z_{\text{out}})}{\epsilon}\right] \tag{5.20}$$

and

$$Y = \mathrm{Im}\left[\frac{z_{\rm in} H_{i\mu}^{(1)\prime}(z_{\rm in})}{\sqrt{\omega_k(0)}} \frac{\sqrt{\omega_k(T)}\bar{H}_{i\mu}^{(1)}(z_{\rm out})}{\epsilon} - \frac{\sqrt{\omega_k(0)}\bar{H}_{i\mu}^{(1)}(z_{\rm in})}{\epsilon} \frac{z_{\rm out}\bar{H}_{i\mu}^{(1)\prime}(z_{\rm out})}{\sqrt{\omega_k(T)}}\right] \tag{5.21}$$

Here $z_{\rm in} = \frac{nM_{KK}}{\epsilon}$ and $z_{\rm out} = \frac{nM_{KK}}{\epsilon}e^{-\epsilon T}$. We can obtain estimates for the energies stored in different momentum modes by using the approximations for Bessel and Hankel functions given in [26]. In particular, we note that for large $z \gg \mu^2 + 1$, we have the following asymptotic expansion

$$H_{i\mu}^{(1)}(z) \approx \sqrt{\frac{2}{\pi z}} e^{\frac{\mu\pi}{2} + i(z - \frac{\pi}{4})} \left[1 + \frac{4\mu^2 + 1}{8iz} + \dots\right] \tag{5.22}$$

To obtain an approximation in the opposite limit, for small z, we recall that $H_{i\mu}^{(1)}(z) = \frac{e^{\mu\pi} J_{i\mu}(z) - J_{-i\mu}(z)}{\sinh \mu\pi}$ and use the fact that

$$J_{i\mu}(z) \approx \frac{e^{i\mu \ln \frac{z}{2}}}{\Gamma(1 + i\mu)} \left[1 - \frac{z^2}{4(1 + i\mu)} + \frac{z^4}{32(1 + i\mu)i\mu} + \dots\right] \tag{5.23}$$

whenever $0 < z \ll \sqrt{|\mu| + 1}$.

For high momentum modes, with $k \to \infty$, we find that $E_k \sim \frac{n^4 M_{KK}^4 \epsilon^2}{4k^5}\left[\cosh^2 \epsilon t - \cos^2 kt\right] e^{-2\epsilon t}$. As expected, these modes are suppressed, being insensitive to the change in the mass of the field. For modes of lower momentum, it is instructive to display the changes in the energy stored in each mode in a characteristic plot, such as the one shown in Fig. 5.1.

The plot shows the energy profile for modes running over a range of different momenta. We also vary over the initial mass of the scalar, nM_{KK}, or equivalently, over the level in the Kaluza-Klein tower. For the φ_n particles, at level n, we see that the total energy density at late times is dominated by modes of momentum $k \lesssim \epsilon$, with significant particle production kicking in at a time $t_n \sim \frac{1}{\epsilon} \ln\left(\frac{nM_{KK}}{\epsilon}\right)$. Using Eqs. (5.22) and (5.23), we can estimate the energy stored in the low energy modes analytically, for $t \gg t_n$. Since $nM_{KK}e^{-\epsilon t} \ll \epsilon \ll nM_{KK}$ and $k \lesssim \epsilon$, we have $z_{\rm out} \ll 1 \ll z_{\rm in}$ and $\mu \lesssim 1$, so that the energies approximate as

$$E_k \approx \frac{ke^{-\frac{k\pi}{\epsilon}}}{2\sinh\left(\frac{k\pi}{\epsilon}\right)} \tag{5.24}$$

In deriving this, we have used the relation $|\Gamma(i\mu)|^2 = \pi/\mu \sinh \mu\pi$, which follows from Euler's reflection formula and the fact that $\Gamma(-i\mu) = \overline{\Gamma(i\mu)}$. To obtain an estimate for the total energy density stored in φ_n particles at late times ($t \gg t_n$), we simply integrate this result over all momentum. The result is

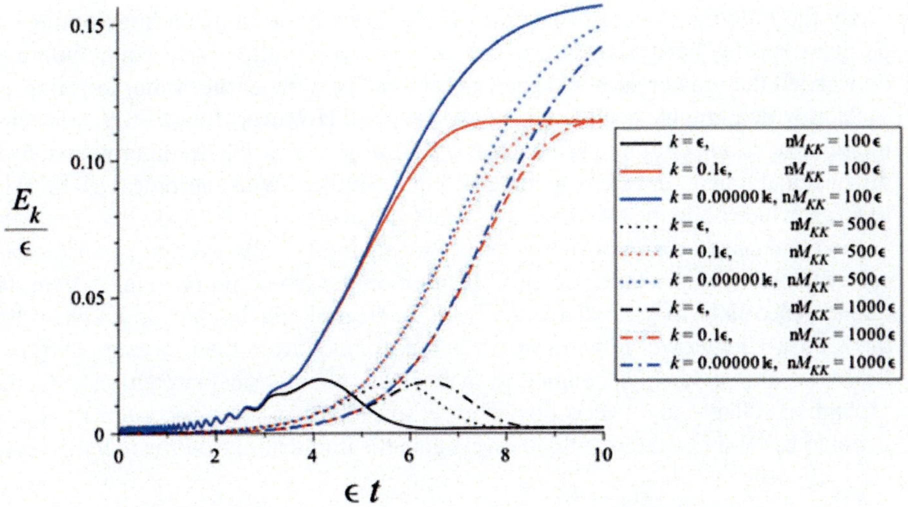

Fig. 5.1 Plot of energy versus time for a range of scalar modes of different momentum and different values for the initial mass. These particular plots were produced using Maple 2020 and a numerical value of $\epsilon = 0.01$

$$\rho_n = \int d^3\mathbf{k} E_k \approx \frac{\pi}{60}\epsilon^4 \tag{5.25}$$

By itself, this would only have a tiny effect on the cosmological background, since $\epsilon^4 = (\alpha\beta)^4 H_0^4$, far below the scale of the critical density of the universe during the dark energy era, $M_{Pl}^2 H_0^2$. Of course, the formula will receive additional corrections from the fact that the fields are actually propagating on a dynamical cosmological background, as opposed to Minkowski, although these are likely to be similarly suppressed, especially if $\alpha\beta \gtrsim 1$. Nevertheless, by the time we have reached $t_N \sim \frac{1}{\epsilon} \ln\left(\frac{N M_{KK}}{\epsilon}\right)$ for some large N, we have started to produce particles for each of the first N levels in the Kaluza-Klein tower. The *total* energy density of all these particles, in this approximation, is given by

$$\rho_{\text{total}}(t \sim t_N) = \sum_{n=1}^{N} \rho_n \approx \frac{N\pi}{60}\epsilon^4 \tag{5.26}$$

This will start to affect the cosmological dynamics for $N \sim N_{\text{crit}}$ where $N_{\text{crit}} \sim \frac{M_{Pl}^2}{\epsilon^2}$, at a time $t_{\text{crit}} \sim \frac{1}{\epsilon} \ln\left(\frac{N_{\text{crit}} M_{KK}}{\epsilon}\right)$. If we assume $\alpha\beta \sim \mathcal{O}(1)$, then $\epsilon \sim H_0$ and so $N_{\text{crit}} \sim 10^{120}$. Assuming $M_{KK} \lesssim M_{Pl}$, so that $N_{\text{crit}} \gg \sqrt{N_{\text{crit}}} \gtrsim \frac{M_{KK}}{\epsilon}$, we see that $t_{\text{crit}} \lesssim \mathcal{O}(400)/H_0$, confirming our expectation that the dark energy era will not last beyond a few hundred or so Hubble times, before the space begins to decompactify. If we lower the underlying Kaluza-Klein scale, M_{KK}, or consider models with $\alpha\beta > 1$, the dark energy era is cut short even earlier.

By the time dark energy succumbs to the extra dimensions, a huge number of particles species have already entered the low energy effective theory. We might be worried that this creates a "species problem" at some earlier time, lowering the scale at which gravity becomes strongly coupled. However, for $N < N_{\rm crit}$ species, the scale of strong coupling scales as $\Lambda_{QG} \sim M_{Pl}/\sqrt{N} > H_0$. In other words, four-dimensional gravity does not become strongly coupled on cosmological scales prior to $t_{\rm crit}$, at which point the effective four-dimensional description breaks down anyway.

As emphasized earlier, everything we are saying is only relevant to cosmological dynamics. On shorter scales, for example in the lab or in the solar system, the quintessence field may be displaced from its cosmological value, so much so that the Kaluza-Klein tower remains heavy and decoupled from the low energy physics. Indeed, if quintessence is coupled to matter with gravitational strength, such a displacement is likely to be necessary in order to avoid fifth force constraints [27–29]. It would certainly be interesting to investigate the implications of this in more detail.

5.3 Further Discussion

In this chapter, we have argued that string theory compactifications consistent with swampland constraints automatically prevent the dark energy era from extending beyond a few hundred Hubble times. This alleviates the coincidence problem to some degree, inasmuch as we now have as much as a percentage chance of finding ourselves in the first e-fold of acceleration. Our conclusions draw on two key features of string compactifications: (i) the scarcity (absence?) of stable de Sitter vacua, suggesting that most dark energy models will be driven by a quintessence field in slow roll [15–17]; (ii) the accumulation of a large number light states as the field rolls off towards infinity [2–5]. Generically, the dynamics is such that the descending tower of light states induces a cosmological phase transition, bringing the dark energy era to a conclusion. We were able to demonstrate this explicitly with a toy model, whereby particle creation in the Kaluza-Klein sector starts to overwhelm the cosmological background within a few hundred Hubble times.[4]

From this stringy perspective, it seems that the cosmological coincidence is readily reduced to one part in $\mathcal{O}(100)$ but can we do any better? Can we exploit the distance conjecture to bring the odds even further in? One speculative possibility is to consider dark energy a consequence of the clockwork dynamics of Chap. 4. The clockwork set-up is brittle. After the dark energy field has moved just a single Planck unit, we could imagine a tower of states contaminating the low scale dynamics and spoiling the clockwork symmetry. Without the clockwork, the dark energy dynamics should also be spoilt. Whilst this idea is appealing, to be able to say anything with any

[4] We note that [30] mentions a similar idea regarding the application of Swampland constraints to the coincidence problem. There, the authors claim that the trans-Planckian censorship conjecture (TCC) fixes the critical time to $\mathcal{O}(H_0)$. However, the analysis of [31] seems to point towards either the TCC being incorrect or not being able to bound de Sitter lifetimes as claimed in [30].

degree of certainty, we would require a clockwork embedding in supergravity, which we have not been able to achieve so far.

Finally, it is natural to ask what would happen if we applied the same analysis to early universe inflation at a much higher scale than quintessence. Because the Kaluza-Klein tower does not need to descend as far to contaminate the inflationary background, the inflaton excursion is limited to just a few Planck units. As a result, we would conclude that, generically, inflation should not extend beyond a few efolds. This seems problematic since inflation must last for at least 50 efolds in order to address the horizon problem. Of course, these concerns aren't new - it is well known that early universe inflation is in some tension with the swampland constraints.

References

1. Cunillera F, Padilla A (2021) A stringy perspective on the coincidence problem. JHEP 10:055. https://doi.org/10.1007/JHEP10(2021)055, arxiv: abs/2105.03426
2. Ooguri H, Vafa C (2007) On the geometry of the string landscape and the swampland. Nucl Phys B 766:21. https://doi.org/10.1016/j.nuclphysb.2006.10.033, arxiv: abs/hep-th/0605264hep-th/0605264
3. Klaewer D, Palti E (2017) Super-planckian spatial field variations and quantum gravity. JHEP 01:088. https://doi.org/10.1007/JHEP01(2017)088, arxiv: abs/1610.00010
4. Grimm TW, Palti E, Valenzuela I (2018) Infinite distances in field space and massless towers of states. JHEP 08:143. https://doi.org/10.1007/JHEP08(2018)143, arxiv: abs/1802.08264
5. Blumenhagen R, Kläwer D, Schlechter L, Wolf F (2018) The refined swampland distance conjecture in Calabi-Yau Moduli spaces. JHEP 06:052. https://doi.org/10.1007/JHEP06(2018)052, arxiv:abs/1803.04989
6. Baume F, Palti E (2016) Backreacted axion field ranges in string theory. JHEP 08:043. https://doi.org/10.1007/JHEP08(2016)043, arxiv:abs/1602.06517
7. Hebecker A, Henkenjohann P, Witkowski LT (2017) Flat monodromies and a Moduli space size conjecture. JHEP 12:033. https://doi.org/10.1007/JHEP12(2017)033, arxiv:abs/1708.06761 1708.06761
8. Cicoli M, Ciupke D, Mayrhofer C, Shukla P (2018) A geometrical upper bound on the inflaton range. JHEP 05:001. https://doi.org/10.1007/JHEP05(2018)001, arxiv:abs/1801.05434 1801.05434
9. Blumenhagen R, Valenzuela I, Wolf F (2017) The swampland conjecture and F-term axion monodromy inflation. JHEP 07:145. https://doi.org/10.1007/JHEP07(2017)145, arxiv:abs/1703.05776 1703.05776
10. Heidenreich B, Reece M, Rudelius T (2018) Emergence of weak coupling at large distance in quantum gravity. Phys Rev Lett 121:051601. https://doi.org/10.1103/PhysRevLett.121.051601, arxiv:abs/1802.08698
11. Landete A, Shiu G (2018) Mass hierarchies and dynamical field range. Phys Rev D 98:066012. https://doi.org/10.1103/PhysRevD.98.066012, arxiv:abs/1806.01874
12. Lee S-J, Lerche W, Weigand T (2018) Tensionless strings and the weak gravity conjecture. JHEP 10:164. https://doi.org/10.1007/JHEP10(2018)164, arxiv:abs/1808.05958 1808.05958
13. Blumenhagen R (2013) Large field inflation/Quintessence and the refined swampland distance conjecture. PoS CORFU2017:175. https://doi.org/10.22323/1.318.0175, arxiv:abs/1804.10504
14. Font A, Herráez A, Ibáñez LE (2019) The swampland distance conjecture and towers of tensionless branes. JHEP 08:044. https://doi.org/10.1007/JHEP08(2019)044, arxiv:abs/1904.05379

15. Obied G, Ooguri H, Spodyneiko L, Vafa C, De Sitter Space and the Swampland. arxiv:abs/1806.08362
16. Garg SK, Krishnan C (2019) Bounds on slow roll and the de Sitter swampland. JHEP 11:075. https://doi.org/10.1007/JHEP11(2019)075, arxiv:abs/1807.05193
17. Ooguri H, Palti E, Shiu G, Vafa C (2019) Distance and de Sitter conjectures on the swampland. Phys Lett B 788:180. https://doi.org/10.1016/j.physletb.2018.11.018, arxiv:abs/1810.05506
18. Andriot D (2019) New constraints on classical de Sitter: flirting with the swampland. Fortsch Phys 67:1800103. https://doi.org/10.1002/prop.201800103, arxiv:abs/1807.09698
19. Hebecker A, Skrzypek T, Wittner M (2019) The F-term problem and other challenges of stringy quintessence. JHEP 11:134. https://doi.org/10.1007/JHEP11(2019)134, arxiv:abs/1909.08625
20. Cicoli M, Cunillera F, Padilla A, Pedro FG (2022) Quintessence and the swampland: the numerically controlled regime of moduli space. Fortsch Phys 70. https://doi.org/10.1002/prop.202200008, arxiv:abs/2112.10783
21. Banerjee A, Cai H, Heisenberg L, Colgáin EO, Sheikh-Jabbari MM, Yang T (2021) Hubble sinks in the low-redshift swampland. Phys Rev D 103:L081305. https://doi.org/10.1103/PhysRevD.103.L081305, arxiv:abs/2006.00244
22. Agrawal P, Obied G, Steinhardt PJ, Vafa C (2018) On the cosmological implications of the string swampland. Phys Lett B 784:271. https://doi.org/10.1016/j.physletb.2018.07.040, arxiv:abs/1806.09718
23. Akrami Y, Kallosh R, Linde A, Vardanyan V (2019) The landscape, the swampland and the era of precision cosmology. Fortsch Phys 67:1800075. https://doi.org/10.1002/prop.201800075, arxiv:abs/1808.09440
24. Copeland EJ, Sami M, Tsujikawa S (2006) Dynamics of dark energy. Int J Mod Phys D 15:1753. https://doi.org/10.1142/S021827180600942X, arxiv:abs/hep-th/0603057hep-th/0603057
25. Mukhanov V, Winitzki S (2007) Introduction to quantum effects in gravity. Cambridge University Press, 6
26. Abramowitz M, Stegun I (1965) Handbook of mathematical functions: with formulas, graphs, and mathematical tables. Dover Publications, Applied Mathematics Series
27. Khoury J, Weltman A (2004) Chameleon fields: awaiting surprises for tests of gravity in space. Phys Rev Lett 93:171104. https://doi.org/10.1103/PhysRevLett.93.171104, arxiv:abs/astro-ph/0309300astro-ph/0309300
28. Khoury J, Weltman A, Chameleon cosmology. Phys Rev D 69:044026. https://doi.org/10.1103/PhysRevD.69.044026, arxiv:abs/astro-ph/0309411astro-ph/0309411
29. Padilla A, Platts E, Stefanyszyn D, Walters A, Weltman A, Wilson T (2016) How to avoid a swift kick in the chameleons. JCAP 03:058. https://doi.org/10.1088/1475-7516/2016/03/058, arxiv:abs/1511.05761
30. Bedroya A, Vafa C (2020) Trans-planckian censorship and the swampland. JHEP 09:123. https://doi.org/10.1007/JHEP09(2020)123 arxiv:abs/1909.11063
31. Dine M, Yu Y, Comments on the transplanckian censorship conjecture. arxiv:abs/2110.07300

Chapter 6
Intermezzo: Supergravity and the Swampland

In the first half of this thesis, we have discussed stringy motivated models and EFTs without making explicit a UV completion. For a consistent EFT treatment, this has the benefit of allowing for concrete and analytical treatment of the four-dimensional low-energy effective field theory by decoupling the microphysics of gravity.

However, this decoupling comes at a price. Without any intuition for what the microphysics are doing, it is hard to argue that the EFT is always well-posed. For example, consider the cosmological constant, a super-IR observable that we might expect to be UV independent. We have seen that the value of the cosmological constant, given by the minimum of the four-dimensional scalar potential, in a string compactification is directly affected by the stabilisation of the moduli fields, which are UV degrees of freedom. Arbitrarily exploring the parameter space of the EFT could, in principle, displace the moduli fields and change the value of the cosmological constant. In extreme cases, this displacement could lead to full destabilisation of the theory and to a break down of the validity of the EFT. Strikingly, this destabilisation might not look pathological on the EFT side and only become a clear problem when considering the full ten-dimensional theory. This kind of heuristic argument about computational control is what motivates, at least in part, the Swampland programme. If true, some of the Swampland arguments, like those in Sect. 2.5.2, can become truly problematic. Indeed, in Sect. 6.2 we show that we are unable to produce a slow-roll regime in the asymptotics of moduli space, implying that there is a disconnect between string theory and cosmology.

In this chapter, we attempt to make more concrete statements regarding lack of computational control and to what extent we can extract phenomenological information from the Swampland programme, that are most relevant to the content of this thesis. A number of authors have explored ideas in the same vein, for example [1] studies the obstacles to generating de Sitter vacua in string theory and its relations with the Swampland and [2] tries to study the formal validity of some common Swampland related lore.

F. Cunillera, *Dark Energy*, Springer Theses,
https://doi.org/10.1007/978-3-031-21468-4_6

6.1 A Simple Example of the Backreaction on Moduli Space

Let us exemplify this situation by studying one of the models in [3]. The type IIB toy model of interest consists of only a Kähler modulus $T = \tau + i\theta$ and the axio-dilaton $S = s + i\sigma$ with a Kähler potential and superpotential given by

$$K = -\log(S + \bar{S}) - 3\log(T + \bar{T})\ , \qquad W = -if + ihS + iqT\ , \tag{6.1}$$

where we note that f is a non-geometric flux which, in [3], is needed for moduli stabilisation purposes. Although there are a vast number of models out there with more complicated compactifications that we may argue are under better control, we choose to concentrate on this simple example since it contains the minimal set of ingredients necessary to see the backreaction of large field excursion onto the theory in an analytical manner.

The scalar potential for the four-dimensional effective theory is calculated from Eq. (2.66) and given by Blumenhagen et al. [3]

$$V = \frac{(hs + f)^2}{16s\tau^3} - \frac{6hqs - 2qf}{16s\tau^2} - \frac{5q^2}{48s\tau} + \frac{\phi^2}{16s\tau^3} \tag{6.2}$$

where $\phi := q\theta + h\sigma$. We are interested in a non-supersymmetric minimum for the scalar potential. The non-SUSY minimum is fixed by

$$\partial_I V = 0 \tag{6.3}$$

where I runs over the different moduli fields. Solving this set of equations, we find a non-SUSY AdS minimum[1]

$$s_0 = \frac{f}{h} \qquad \tau_0 = \frac{6f}{5q} \qquad \phi_0 := q\theta_0 + h\sigma_0 = 0 \qquad V_0 = -\frac{25hq^3}{216f^2} \tag{6.4}$$

with a cosmological constant V_0. We would like to consider ϕ as our candidate inflaton field and see what is the effect of the back-reaction of the rolling inflaton field. We are interested on the moduli dependence with ϕ when we displace it from its minimum ϕ_0. To find $s(\phi)$ and $\tau(\phi)$, we relax Eq. (6.3) and solve

$$\begin{cases} \partial_s V = 3h^2s^2 + 5q^2\tau^2 - 6fq\tau - 3f^2 - 3\phi^2 = 0 \\ \partial_\tau V = 5q^2\tau^2 + 12q(3hs - f)\tau - 9(hs + f)^2 - 9\phi^2 = 0 \end{cases}, \tag{6.5}$$

while allowing $\partial_\phi V \neq 0$. From here, we find

[1] Note that the orthogonal direction to ϕ remains as a flat axionic direction as it does not appear in the potential.

$$s(\phi) = \frac{f}{h}\sqrt{1+\frac{5}{8}\left(\frac{\phi}{f}\right)^2}\,, \qquad \tau(\phi) = \frac{6f}{10q}\left(1+\sqrt{1+\frac{5}{8}\left(\frac{\phi}{f}\right)^2}\right)\,. \tag{6.6}$$

When ϕ becomes of order $\phi_c = \sqrt{\frac{8}{5}}f$, the back-reaction of the inflaton field is substantial and cannot be ignored. This supports the idea that naive displacements away from the minimum have to be bounded in order to maintain the validity of the EFT description. To derive a physically meaningful quantity, we concentrate on finding the critical value for the canonically normalised inflaton field.

We note that the canonically normalised inflaton is given by

$$\Phi = \int \sqrt{K_{IJ}\frac{\partial m^I}{\partial\phi}\frac{\partial m^J}{\partial\phi}}\,d\phi = \int \left(n_I K^{IJ} n_J\right)^{-\frac{1}{2}}\,d\phi\,, \tag{6.7}$$

where m^I stands for the moduli fields with fluxes n_I

$$\phi = \sum_I n_I m^I\,. \tag{6.8}$$

Plugging back our scenario we find for the subcritical regime ($\phi < \phi_c$)

$$\Phi(s,v) = \sqrt{\frac{3}{3h^2s^2+q^2v^2}}\,\phi\,, \qquad \Phi(s_0,v_0) = \frac{5}{\sqrt{37}}\frac{\phi}{f}\,. \tag{6.9}$$

For this simple model, we can see that there is a critical value for the canonical inflaton $\Phi_c = \sqrt{\frac{40}{37}} \simeq 1.03$ which is flux independent and its value is around one Planck unit, i.e. it cannot be tuned by fluxes.

These results are in agreement with the refined Swampland distance conjecture in Sect. 2.5. Finally, to make contact with the light tower of states we note that

$$M_{\rm KK} \sim \frac{1}{\tau} \sim \frac{q}{f}e^{-\frac{\gamma}{2}\Phi}\,, \tag{6.10}$$

which drop off exponentially fast as claimed by the refined conjecture.

In this example, we have considered a type IIB toy model without complex structure moduli, with a flat direction σ and a non-geometric flux f. Although one might be tempted to criticise this particular model for its simplicity, we find it more enlightening to take it as an example of a larger class of models that belong to the swampland which share some common ideas behind them. We discuss some of these ideas below.

6.2 Calculations at the Boundary of Moduli Space

The strict limit to the boundary of moduli space, thought of as the limit of the dilaton or the volume modulus going to infinity, is a limit to ten-dimensional Minkowski. Given that our Universe is not Minkowski, or ten-dimensional for that matter, this makes drawing conclusions for phenomenology in this limit somewhat useless.

In the following, we will prove that quintessence is not possible in the runaway region arbitrarily close to the boundary of moduli space, because one cannot satisfy the slow roll constraints. Related conclusions were drawn in [4–6], here we build on these works by analysing simultaneously the Kähler-dilaton sectors, by highlighting the role of supersymmetry and by considering corrections to the moduli potentials. Further difficulties with quintessence model building in string theory were also discussed in [7]. In the following we discuss the absence of a slow roll region in the runaway of type II and heterotic supergravities. For a comprehensive review of the pros-and-cons of the different theories regarding other model building aspects see [8].

To proceed with our proof, let us approach the boundary of moduli space in a string theory compactification on some Calabi-Yau threefold, $\mathcal{X}$. In the strict boundary limit, we should think of the dilaton or the volume moduli going to infinity. As is well known, such a limit is expected to bring down an infinite tower of light states, spoiling the effective field theory description [9–12]. In particular, the asymptotic limits of the volume moduli bring down either Kaluza-Klein modes, consistent with a decompactification of the internal space, or winding modes, consistent with decompactification of the dual. For the asymptotic limits of the dilaton, the tower of light states is less clear—it has been suggested that this can introduce a tower of tensionless strings and domain walls [13].

Since the physics at the boundary is clearly not phenomenologically viable, let us relax the limit and consider what happens when the dilaton or some Kähler modulus becomes arbitrarily large but finite. In this case, the effective theory corresponds to tree-level supergravity close to the boundary, where the moduli fields are in a runaway regime. We are interested in theories that might admit no four-dimensional vacuum, or potentially a supersymmetric Minkowski or AdS vacuum somewhere in the bulk of moduli space. The supersymmetric requirement serves to align our discussion with the usual swampland lore of non-supersymmetric vacua being unstable [14]. It follows from the arguments given in [15] that de Sitter vacua cannot arise in any parametrically controlled regime of the moduli space, and our only hope to achieve phenomenologically viable models of dark energy is to consider runaway quintessence. For example, type IIB flux compactifications feature at tree-level supersymmetric Minkowski vacua where the Kähler moduli are flat directions due to the underlying no-scale structure [16]. Leading order α' corrections, or a non-supersymmetric stabilisation of the dilaton and the complex structure moduli, can then generate this type of de Sitter runaway for large volume.

In the following we will assume that there exists some mechanism to fix the complex structure moduli at tree-level in a controlled manner,[2] while the remaining

[2] Or the Kähler moduli in case of the type IIA discussion.

moduli fields correspond to the axio-dilaton $S = s + i\alpha$ and the Kähler moduli $T^a = \tau^a + i\theta^a$, which are identified with either two-cycles or four-cycles, depending on which is most convenient. In the runaway regime, close to the boundary, the dynamics of these moduli is governed by the following Kähler potential in its tree-level sum separable form

$$K = -p\ln(\mathcal{V}) - \ln(S+\bar{S}) + K_0\,, \tag{6.11}$$

and a superpotential W to be specified for each supergravity. We have denoted the internal volume by $\mathcal{V}(\tau^a)$. For any Calabi-Yau threefold $\mathcal{X}$, the volume is a homogeneous function of degree 3/2 in the Kähler moduli for four-cycles, or equivalently, of degree 3 in the Kähler moduli for two-cycles. Furthermore, for type IIB supergravity, we have $p = 2$, and work with four-cycles, while for type IIA and heterotic supergravities, we have $p = 1$, and work with two-cycles. As already stated, the complex structure contribution to the Kähler potential, K_0, is considered to be fixed. For further details, see [17, 18].

The Lagrangian for the scalar moduli is given by

$$\mathcal{L} = K_{S\bar{S}} dS \wedge \star d\bar{S} + K_{a\bar{b}} dT^a \wedge \star dT^{\bar{b}} - V, \tag{6.12}$$

where $K_{I\bar{J}} = \partial_I \partial_{\bar{J}} K$ is the Kähler metric for moduli space, which given equation Eq. (6.11) is block diagonal. The scalar potential is obtained by computing

$$V = e^K (K^{I\bar{J}} D_I W D_{\bar{J}} \bar{W} - 3|W|^2), \tag{6.13}$$

where $D_I W = \partial_I W + W \partial_I K$ is the Kähler covariant derivative, and I runs over all the moduli of the theory.

At the perturbative level, the axions, $\theta^a = \mathrm{Im}\,(T^a)$ and α, do not contribute to the scalar potential and cannot play the rôle of runaway quintessence. Focusing instead on the saxions/moduli, $\tau^a = \mathrm{Re}\,(T^a)$ and s, we find that the relevant part of the Lagrangian is now given by

$$\mathcal{L} = \frac{1}{4s^2}(\partial s)^2 + K_{a\bar{b}} \partial\tau^a \partial\tau^b - V(s, \tau^a)\,, \tag{6.14}$$

A necessary condition for phenomenologically viable quintessence is the existence of a slow roll regime, defined by the condition $\epsilon := -\dot{H}/H^2 < 1$ where H is the Hubble parameter and dot denotes differentiation with respect to proper cosmological time. For a generic multiscalar theory described by a Lagrangian $\mathcal{L} = \frac{1}{2} Z_{IJ} \partial\phi^I \partial\phi^J - V(\phi^I)$ one can show in the limit of vanishing acceleration, $\ddot{\phi}^I = 0$, the dynamics of the system is described by

$$H^2 \approx \frac{1}{3} V \quad \text{and} \quad 3H\dot{\phi}^I + \Gamma^I_{JK} \dot{\phi}^J \dot{\phi}^K \approx -Z^{IJ} \partial_{\phi^J} V, \tag{6.15}$$

where Z^{IJ} is the inverse of the field space metric and Γ^I_{JK} is the corresponding metric connection. These equations admit two distinct classes of slow roll solutions, whose existence and stability has been analysed in the context of dark energy in [19]. In this note we will focus on the cases when $3H\dot{\phi}^I \gg \Gamma^I_{JK}\dot{\phi}^J\dot{\phi}^K \to 0$, for which

$$\dot{H} \approx \frac{(\partial_{\phi^I} V)\,\dot{\phi}^I}{6H} \approx -\frac{(\partial_{\phi^I} V)Z^{IJ}(\partial_{\phi^J} V)}{18H^2}, \tag{6.16}$$

and so

$$\epsilon = -\frac{\dot{H}}{H^2} \approx \frac{1}{2}\left(\partial_{\phi^I} \ln V\right) Z^{IJ} \left(\partial_{\phi^J} \ln V\right). \tag{6.17}$$

In this regime, the requirement of quasi-de Sitter expansion implies the need for flat (multifield) scalar potentials. The other class of accelerating solutions, that we will not analyse here, forego the flatness of the potential in exchange for large field space curvature, see [19, 20].[3] While promising, this avenue is not without challenges from a string model building perspective [22].

For the supergravity compactifications under consideration Eq. (6.14), the first slow roll parameter is given by $\epsilon = \epsilon_s + \epsilon_{\mathcal{V}}$, where the dilaton contribution is

$$\epsilon_s = s^2\left(\partial_s \ln V\right)^2, \tag{6.18}$$

and the Kähler contribution is

$$\epsilon_{\mathcal{V}} = \frac{1}{4}(\partial_{\tau^a} \ln V)K^{a\bar{b}}(\partial_{\tau^b} \ln V), \tag{6.19}$$

Throughout this section, we will make regular use of Euler's theorem and its corollary: if P_n is a homogeneous function of degree n in τ^a, then

$$\tau^a\partial_{\tau^a} \ln P_n = n, \qquad \tau^a\partial_{\tau^a}\partial_{\tau^b} \ln P_n = -\partial_{\tau^b} \ln P_n.$$

In particular, this allows us to infer the following conditions on the derivatives of the Kähler potential

$$K^{a\bar{b}}\partial_{\bar{T}^b}K = -2\tau^a, \qquad (\partial_{T^a}K)K^{a\bar{b}}(\partial_{\bar{T}^b}K) = 3, \tag{6.20}$$

where the indices run over the Kähler moduli only. The latter result holds as long as we work with four-cycle Kähler moduli in type IIB supergravity, and two-cycle Kähler moduli in type IIA and heterotic.

Let us now derive the detailed form of the tree-level scalar potential, and the corresponding slow roll parameter, for the type II and heterotic supergravities, in turn. We will see that the slow roll condition $\epsilon < 1$ can never be satisfied in the runaway regions close to the boundary of moduli space.

---

[3] Strictly speaking the requirement is that $\Gamma^I_{JK}\dot{\phi}^J\dot{\phi}^K \gg 3H\dot{\phi}^I$ for one of the scalars, which can be achieved even in flat field space if, for instance, one uses polar coordinates as in [21].

6.2.1 The Type IIB Runaway

Let us begin with type IIB supergravity (for further details, see [17]). In this instance, at tree-level, the superpotential is linear in the dilaton $W = h_0 S + f_0$, with h_0 and f_0 set by the three-form fluxes, respectively H_3 and F_3, that stabilise the complex structure moduli supersymmetrically. Furthermore, the Kähler moduli $T^a = \tau^a + i\theta^a$ that enter the Kähler potential through the volume will be identified with four-cycles, so that the volume is a homogeneous function of degree $\frac{3}{2}$ in the corresponding saxions, τ^a. Using the form of the F-term scalar potential given by equation Eq. (6.13) and the Kähler metric Eq. (6.11) with $p = 2$, we obtain a scalar potential

$$V = \frac{e^{K_0}}{2s\mathcal{V}^2}|h_0\bar{S} - f_0|^2 , \tag{6.21}$$

where $\bar{S} = s - i\alpha$. If $h_0 \neq 0$, the imaginary part of the axio-dilaton can be stabilised at the point where $\partial_\alpha V = 0$, or equivalently, $\alpha = -\text{Im}\left(\frac{f_0}{h_0}\right)$, so that the scalar potential becomes

$$V(s, \mathcal{V}) = \frac{e^{K_0}}{2s\mathcal{V}^2}|h_0|^2 \left(s - \text{Re}\left(\frac{f_0}{h_0}\right)\right)^2 . \tag{6.22}$$

It follows that

$$\epsilon_s = \left(\frac{s + \text{Re}\left(\frac{f_0}{h_0}\right)}{s - \text{Re}\left(\frac{f_0}{h_0}\right)}\right)^2 , \qquad \epsilon_{\mathcal{V}} = 3, \tag{6.23}$$

defined whenever $V \neq 0$. We immediately infer that $\epsilon = \epsilon_s + \epsilon_{\mathcal{V}} \geq 3$, ruling out slow roll in the runaway regime. Notice that vacua with $h_0 = 0$ feature exactly $\epsilon_s = 1 \Rightarrow \epsilon = 4$.

For type IIB supergravity, if $h_0 \neq 0$, it is possible to further stabilise the dilaton at the location of the supersymmetric minimum, $s = \text{Re}\left(\frac{f_0}{h_0}\right)$, where the leading order potential for the Kähler moduli vanishes. When this happens, we should consider additional perturbative and non-perturbative corrections to the potential. In order to be compatible with a robust large volume expansion, these go as [16, 23, 24]

$$V = \frac{\mathcal{A}}{\mathcal{V}^{2+p}} + \frac{\mathcal{B}e^{-f}}{\mathcal{V}^{2+q}} + \frac{\mathcal{C}}{\mathcal{V}^{2+r}g^n} \dots , \tag{6.24}$$

where $\mathcal{A}$, $\mathcal{B}$ and $\mathcal{C}$ will in general depend on the stabilised dilaton and complex structure moduli, while $f > 0$ and $g > 0$ are homogeneous functions of degree one in the τ^a. In particular, f corresponds to the dominant saxion in the non-perturbative expansion. In order for us to trust the large volume expansion, we require each of these terms to scale away more quickly than the leading order $1/\mathcal{V}^2$ term, which happened to vanish once the dilaton was stabilised at its supersymmetric minimum.

This implies that $p > 0, q \geq 0$, and $r + \frac{2n}{3} > 0$, suggesting that the volume direction is made even steeper and the volume modulus will not slow roll in this case either. To see this explicitly, we compute the slow roll parameter in the asymptotic regime where only one of the terms dominates, in accordance with our definition of the boundary of moduli space. If $V \sim \frac{\mathcal{A}}{\mathcal{V}^{2+p}}$, we find that $\epsilon \sim 3\left(1+\frac{p}{2}\right)^2 > 3$. In contrast, if $V \sim \frac{\mathcal{B}e^{-f}}{\mathcal{V}^{2+q}}$, we find that

$$\epsilon = 3\left(1+\frac{q}{2}\right)^2 + f(2+q) + \frac{1}{4}K^{a\bar{b}} f_a f_b \geq 3\,, \tag{6.25}$$

where the inequality follows from the fact that $f, q \geq 0$ and $K^{a\bar{b}}$ being a symmetric matrix with all positive eigenvalues. Finally, if $V \sim \frac{\mathcal{C}}{\mathcal{V}^{2+r} g^n}$ we find that

$$\epsilon = 3\left(1+\frac{r}{2}\right)^2 + n(2+r) + \frac{n^2}{4}K^{a\bar{b}}(\ln g)_a(\ln g)_b \geq 3\,. \tag{6.26}$$

In each case we see that slow roll is impossible in this asymptotic region, where a single one of the corrections in equation Eq. (6.24) dominates. It is interesting to note that leading order supersymmetry, imposed for the sake of the stability of the compactification, only aggravates the problem, by cancelling off the $1/\mathcal{V}^2$ term that would otherwise dominate the potential at the boundary of moduli space and replacing it with a steeper term.

Alternatively, we could consider a non-supersymmetric stabilisation of the complex structure sector. This would induce a correction to Eq. (6.22) of the form $\lambda/s\mathcal{V}^2$, where λ is a positive constant proportional to the F-terms of the complex structure moduli. The vacuum expectation value of the dilaton is then shifted to a non-supersymmetric value leaving a $1/\mathcal{V}^2$ runaway for the volume mode, which is too steep to give rise to slow roll. Note further that if we move into the bulk of moduli space, we might hope to stabilise the volume at some fixed value and achieve slow roll along some other saxion direction, through the last term in Eq. (6.24). This is indeed possible, although such a scenario runs into further difficulties associated with a light gravitino [25]. It is also possible that interference between two terms in equation Eq. (6.24) can give rise to a (short) field range for the volume where $\epsilon < 1$. This avenue is likely to involve considerable tuning and, by definition, lies in the bulk of moduli space and therefore will not be analysed in this chapter.

6.2.2 *The Type IIA Runaway*

We now turn our attention to the type IIA runaway (for further details, see [17]). The superpotential now contains RR fluxes (e_0, e_a, q^a), the (3,0)-component of the H_3 flux, h_0, and the Romans mass, m [26]:

$$W = e_0 + e_a T^a + \frac{1}{2}\mathcal{K}_{abc} q^a T^b T^c + \frac{m}{6}\mathcal{K}_{abc} T^a T^b T^c - i h_0 S \,. \tag{6.27}$$

Here the Kähler moduli $T^a = v^a + i\theta^a$ are identified with two-cycles and the volume is a homogeneous function of degree 3 in the corresponding saxions, v^a. In particular, $\mathcal{V} = \frac{1}{6}\mathcal{K}_{abc} v^a v^b v^c$, where $\mathcal{K}_{abc}$ are the triple intersection numbers for the Calabi-Yau. Without additional fluxes or corrections to the tree-level action, the complex structure sector remains flat. To continue the discussion on the slow roll regime, we must therefore assume that some other mechanism under computational control exists to stabilise the complex structure sector, so that, at tree-level, it only enters through some constant in the Kähler potential, $K_0 = \int_{\mathcal{X}} \Omega \wedge \bar{\Omega}$.

The tree-level scalar potential for the type IIA theory is then given by

$$V = \frac{e^{K_0}}{2s\mathcal{V}} \left[4|h_0|^2 s^2 + K^{a\bar{b}} \rho_a \bar{\rho}_b + 4s\mathrm{Im}(W\bar{h}_0) - 4\tau^a \mathrm{Re}(W\bar{\rho}_a) + |W|^2 \right] \,, \tag{6.28}$$

where

$$\rho_a = \partial_{T^a} W = e_a + \mathcal{K}_{abc} q^b T^c + \frac{m}{2}\mathcal{K}_{abc} T^b T^c \,. \tag{6.29}$$

We can always absorb the vacuum expectation value of the axions into a redefinition of the fluxes. Therefore, without loss of generality, we set the axions to vanish. With the axions gone, the superpotential can be written as

$$W = e_0 + e_a v^a + \mathcal{V}(q^a \omega_a + m) - i h_0 s \,, \tag{6.30}$$

and

$$\rho_a = e_a + \mathcal{V}\left[q^b(\omega_{ab} + \omega_a \omega_b) + m\omega_a \right] \,, \tag{6.31}$$

where $\omega_a = \partial_{v^a} \ln \mathcal{V}$ and $\omega_{ab} = \partial_{v^a}\partial_{v^b} \ln \mathcal{V}$. Keeping only the leading order Kähler terms at large volume, we obtain a scalar potential that goes as

$$V = \frac{e^{K_0}}{2s\mathcal{V}} \left[|h_0|^2 s^2 + 14\mathrm{Im}(m\bar{h}_0)\mathcal{V}s + |m|^2 \mathcal{V}^2 \right] + \ldots \tag{6.32}$$

We now compute the slow roll parameter. Since the scalar potential is a function of $\mathcal{V}/s$ to leading order in this asymptotic regime, we have that $\epsilon_{\mathcal{V}} = 3\epsilon_s$. Furthermore,

$$\epsilon_s = \left[\frac{|h_0|^2 - |m|^2 \frac{\mathcal{V}^2}{s^2}}{|h_0|^2 + 14\mathrm{Im}(m\bar{h}_0)\frac{\mathcal{V}}{s} + |m|^2 \frac{\mathcal{V}^2}{s^2}} \right]^2 \,. \tag{6.33}$$

We remark that, in order to keep α' corrections from becoming important, we require[4]

$$\mathcal{V}s \gg 1 \,, \tag{6.34}$$

[4] In units of $\alpha' = 1$.

where $\mathcal{V}_S$ is the volume of $\mathcal{X}$ in the string frame. After expressing the Einstein frame volume in string frame, $\mathcal{V} = s^{3/2}\mathcal{V}_S$, we see that $\mathcal{V}/s = \sqrt{s}\mathcal{V}_S \gg 1$ to be consistent with Eq. (6.34) and weak coupling. Thus, in the runaway regime, the ratio $\mathcal{V}/s$ has to be very large and it follows that $\epsilon_s \approx 1$, and so $\epsilon = \epsilon_s + \epsilon_{\mathcal{V}} \approx 4$, ruling out slow roll.

6.2.3 *The Heterotic Runaway*

We finish with the heterotic runaway, where the tree-level superpotential does not depend on the dilaton or the Kähler moduli (for further details see [18]), so that $W = W_{cs}$. The complex structure moduli are assumed to lie in their supersymmetric vacuum by the vanishing of the corresponding F-terms. As a result, the scalar potential becomes a runaway in the dilaton and the volume modulus

$$V = e^{K_0} \frac{|W_{cs}|^2}{2s\mathcal{V}} \,. \tag{6.35}$$

Computing the slow roll parameters, we find that $\epsilon_s = 1$ and $\epsilon_{\mathcal{V}} = 3$, and so $\epsilon = 4$. Clearly slow roll cannot be achieved in the heterotic runaway.

We have found that the tree-level type II and heterotic supergravities cannot contain a slow roll region in a parametrically controlled regime. The result is proven for any number of Kähler moduli. It is clear that obtaining a slow roll region requires breaking the form of the Kähler potential in Eq. (6.11). This can be done by introducing corrections to the tree-level action. In Chap. 8, we will step away from the runaway, by considering perturbative and non-perturbative corrections to this leading order behaviour in detail. There we find that phenomenologically viable quintessence requires non-supersymmetric vacua. This suggests that proponents of the swampland should now object to quintessence as vigorously as they object to de Sitter, placing them on a collision course with observations.

6.3 Divergence of String Diagrams and Perturbative Control or Lack Thereof

A statement against the computational control of theory away from the boundary space is that, for finite couplings, the string diagrams can be written as an expansion in these couplings which is formally divergent when summed over. We do not seem to be able to find the origin of this statement for the Swampland programme, but a qualitatively similar argument exist for QED.[5]

---

[5] We thank Tony Padilla for pointing out this analogy.

Indeed, in [27] it was already pointed out that the power series in e^2 that one obtains as a results of the perturbative treatment of QED are divergent, even though after renormalisation each individual coefficient in the series is finite. The argument given by Dyson is beautiful in its simplicity and we review it below. Let

$$F(e^2) = a_0 + a_2 e^2 + e_4 e^4 + ... \qquad (6.36)$$

be a physical quantity that admits a power expansion in e^2. Let $F(e^2)$ converge for some positive value of e^2, then the series must be an analytic function around $e = 0$. By analytical continuation, one could then consider small enough values of $-e^2$ such that the series is also a well-behaved analytic function with a convergent expansion. Dyson thinks of this as a fictitious world where elementary particles of equal charge attract.

In this fictitious world, the Hamiltonian is not bounded from below, and there is no good definition of a vacuum state. Indeed, the state of minimal energy will be occupied by an infinite number of these $-e^2$-charged states. Even if only a finite number of these states is initially present in the theory, spontaneous polarization of the would-be lowest-energy state would rapidly swamp the theory again. In this precarious situation, it is not possible to start with any given state in the fictitious world and, after integrating the equations of motion over some finite (or infinite) time, end up with well-defined analytic functions. Thus $F(-e^2)$ cannot be analytic and the series Eq. (6.36) must diverge. This is true for any $e^2 \neq 0$; of course if the electron charge were to vanish the sum would be trivially convergent.

However, nearly 70 years after the publication of [27] we know that QED is one of the most precise theories available to physicist. One of its many successes is the theoretical calculation of the anomalous magnetic moment of the electron a_e. The QED contribution to a_e can be written as an expansion in Schwinger factors to n-loops as

$$a_e^{(\mathrm{QED})} = \sum_{n=1}^{\infty} \left(\frac{\alpha}{\pi}\right)^n a_e^{(2n)} , \qquad (6.37)$$

where $a_e^{(2n)}$ is some finite quantity. This sum has been calculated to 5-loops [28] and, in comparison with the experimental value [29], one finds that

$$\left|a_e^{(\mathrm{QED})} - a_e^{(\mathrm{exp})}\right| \sim 10^{-12} . \qquad (6.38)$$

The agreement of QED with Nature is hard to argue against, even though Eq. (6.37) is formally divergent[6]. Already in [27], the author provided an explanation to this uncomfortable problem. Either the theory is complete and we do not have the appropriate mathematical tools to make sense of the perturbative expansion or the theory is incomplete and the perturbative expansion is merely a workable approximation to a complete physical theory.

---

[6] In two different ways, in fact. One by Dyson's argument and two from the multiplicity of Feynman diagrams scaling as $n!$ so that at some point in loop expansion, $n!\alpha^n \sim 1$.

An example that exemplifies these arguments by Dyson is given by the quantum anharmonic oscillator of [30, 31],

$$\left(\frac{d^2}{dx^2} + \frac{x^2}{4} + \frac{\lambda x^4}{4} - E(\lambda)\right)\psi(x) = 0\,. \tag{6.39}$$

This theory can be solved, without making use of any approximation, by numerical techniques. An analytic solution requires making use of a perturbative expansion in the quartic coupling λ such that the energy eigenstates take the form

$$E^k \sim k + \frac{1}{2} + \sum_{n=1}^{\infty} A_n^K \lambda^n\,, \qquad K = 0, 1, 2, \ldots \tag{6.40}$$

The perturbative approach suffers from the same issues as QED. The energy eigenstates have a zero radius of convergence in the λ coupling. However, perturbation theory is able to reproduce the numerical results accurately [32]. Furthermore, the perturbative expansion has been shown to be Borel summable [33], which renders the perturbative theory convergent[7]. We view this as an example of a theory that is closed, has perturbative infinities, and yet remains accurate with respect to the exact numerical results, and where new mathematical techniques give us a better understanding of the perturbative expansion.

Coming back to supergravity, we take Dyson's point of view. Even if there are formal infinities in the theory, as long as the perturbative expansion is well-behaved order by order, we should treat it as an approximation to the true physics of string theory. After all, supergravity only aims to capture the massless sector of string theory, so it was never a complete theory to begin with.

6.3.1 *Perturbative and Non-perturbative Contributions Spoiling the Vacuum*

A number of Swampland conjectures exist to argue that, away from the boundary of moduli space, a number of corrections are bound to appear that spoil the original vacuum. For example, in [34], the authors argue that solutions that admit a leading order supersymmetric Minkowski vacuum, only broken at subleading order by instanton corrections, will exhibit non-perturbative corrections that will destabilise the leading order solution. A number of toroidal examples are given to justify this conjecture.

In a more heuristic sense, this type of Swampland conjectures aim to draw non-model specific statements about string compactifications from a number of examples. These behaviours seem general enough that finding counterexamples is hard, if at all

[7] We thank Oliver Gould for pointing out this nice example.

possible. At the same time, no formal and general proof for these conjectures exist. With regards to the topics in this thesis, conjectures about the existence and stability of de Sitter vacua are most relevant. These are not new or limited to the Swampland programme. We have already discussed these to some extent in Sect. 2.5.2. A very complete rundown of de Sitter and the Swampland can be found in [35, 36], and we will not dwell on this any further.

We would like to finally remark that the question of systematically and consistently expanding supergravity order by order in the perturbative expansion of its coupling is an ongoing area of research. As we have alluded before, showing that the effective theory is not spoilt by obscure corrections is highly non-trivial due to supergravity having two expansion parameters. Some work in this direction has been done in [24], where the authors were able to show that the leading no-scale breaking terms appear at $\mathcal{O}(\alpha'^3)$ in the four-dimensional effective theory, modulo potential logarithmic corrections. Another work [37], showed that the ten-dimensional equations of motion and the four-dimensional equations of motion for KKLT [38] are equivalent while also providing insight into the cancellation of divergences at the ten-dimensional level.

We conclude that, Swampland programme aside, a number of interesting questions remain open. One is to complete the work of [24] in showing that no-scale can be consistently broken at $\mathcal{O}(\alpha'^3)$, *i.e.* the leading order vacuum generated by instanton and quartic corrections to the ten-dimensional Ricci scalar $\mathcal{R}^4_{(10)}$ is safe from instability woes. A second one is to better understand these expansions in terms of ten-dimensional physics, extending the work of [37] to non-KKLT models.

We have seen that studying cosmology at the boundary of moduli space seems highly implausible, at best. In most of the remainder of our thesis we will shift our attention to challenges of model building our Universe in the bulk of moduli space. We will see that a hierarchy of scales between different orders in perturbation theory is necessary to stabilise the moduli fields. We argue that this scale hierarchy is precisely the one that protects the perturbative expansion against unwanted instabilities. Of course, this philosophy is not new but rather one of the usual sanity checks that one employs when constructing supergravity models at large (but finite) volume and small (but finite) coupling.

In the next chapter, we will introduce the necessary perturbative and non-perturbative tools to break the tree-level no-scale structure and generate a minimum for the Kähler sector, thereby stabilising all the moduli fields in the theory. In Chap. 8 we explore the challenges of building quintessence from string theory and provide a "series of ingredients" necessary that any phenomenologically viable quintessence model must contain. We will find that an understanding of next-to-leading order contributions can become important for model builders. Finally, in Chap. 9 we begin to think in this direction and study whether higher curvature corrections can help produce de Sitter vacua.

References

1. Dine M, Law-Smith JAP, Sun S, Wood D, Yu Y (2021) Obstacles to constructing de Sitter space in string theory. JHEP 2:50. https://doi.org/10.1007/JHEP02(2021)050. https://arxiv.org/abs/2008.12399
2. Andriot D (2019) New constraints on classical de Sitter: flirting with the swampland. Fortsch Phys 67:1800103. https://doi.org/10.1002/prop.201800103. https://arxiv.org/abs/1807.09698
3. Blumenhagen R, Valenzuela I, Wolf F (2017) The swampland conjecture and F-term axion monodromy inflation. JHEP 07:145. https://doi.org/10.1007/JHEP07(2017)145. https://arxiv.org/abs/1703.05776
4. Hertzberg MP, Kachru S, Taylor W, Tegmark M (2007) Inflationary constraints on type IIA string theory. JHEP 12:095. https://doi.org/10.1088/1126-6708/2007/12/095. https://arxiv.org/abs/0711.2512
5. Garg SK, Krishnan C, Zaid Zaz M (2019). Bounds on slow roll at the boundary of the landscape. JHEP 03:029. https://doi.org/10.1007/JHEP03(2019)029, https://arxiv.org/abs/1810.09406
6. Valeixo Bento B, Chakraborty D, Parameswaran SL, Zavala I (2020) Dark energy in string theory. PoS CORFU2019 123. https://doi.org/10.22323/1.376.0123, https://arxiv.org/abs/2005.10168
7. Hebecker A, Skrzypek T, Wittner M (2019) The F-term problem and other challenges of stringy quintessence. JHEP 11:134. https://doi.org/10.1007/JHEP11(2019)134, https://arxiv.org/abs/1909.08625
8. Cicoli M, De Alwis S, Maharana A, Muia F, Quevedo F (2019) De Sitter vs quintessence in string theory. Fortsch Phys 67:1800079 https://doi.org/10.1002/prop.201800079, https://arxiv.org/abs/1808.08967
9. Ooguri H, Vafa C (2007) On the geometry of the string landscape and the swampland. Nucl Phys B 766:21. https://doi.org/10.1016/j.nuclphysb.2006.10.033, https://arxiv.org/abs/hep-th/0605264
10. Klaewer D, Palti E (2017) Super-Planckian spatial field variations and quantum gravity. JHEP 01:088. https://doi.org/10.1007/JHEP01(2017)088, https://arxiv.org/abs/1610.00010
11. Grimm TW, Palti E, Valenzuela I (2018) Infinite distances in field space and massless towers of states. JHEP 08:143. https://doi.org/10.1007/JHEP08(2018)143, https://arxiv.org/abs/1802.08264
12. Blumenhagen R, Kläwer D, Schlechter L, Wolf F (2018) The refined swampland distance conjecture in Calabi-Yau moduli spaces. JHEP 06:052. https://doi.org/10.1007/JHEP06(2018)052, https://arxiv.org/abs/1803.04989
13. Font A, Herráez A, Ibáñez LE (2019) The swampland distance conjecture and towers of tensionless branes. JHEP 08:044. https://doi.org/10.1007/JHEP08(2019)044, https://arxiv.org/abs/1904.05379
14. Ooguri H, Vafa C (2017) Non-supersymmetric AdS and the Swampland. Adv Theor Math Phys 21:1787. https://doi.org/10.4310/ATMP.2017.v21.n7.a8, https://arxiv.org/abs/1610.01533
15. Ooguri H, Palti E, Shiu G, Vafa C (2019) Distance and de Sitter conjectures on the swampland. Phys Lett B 788:180. https://doi.org/10.1016/j.physletb.2018.11.018, https://arxiv.org/abs/1810.05506
16. Burgess CP, Cicoli M, Ciupke D, Krippendorf S, Quevedo F (2020) UV shadows in EFTs: accidental symmetries, robustness and no-scale supergravity. Fortsch Phys 68:2000076. https://doi.org/10.1002/prop.202000076, https://arxiv.org/abs/2006.06694
17. Grimm TW (2005) The effective action of type II Calabi-Yau orientifolds. Fortsch Phys 53:1179. https://doi.org/10.1002/prop.200510253, https://arxiv.org/abs/hep-th/0507153
18. Cicoli M, de Alwis S, Westphal A (2013) Heterotic moduli stabilisation. JHEP 10:199. https://doi.org/10.1007/JHEP10(2013)199, https://arxiv.org/abs/1304.1809
19. Cicoli M, Dibitetto G, Pedro FG (2020) New accelerating solutions in late-time cosmology. Phys Rev D 101:103524. https://doi.org/10.1103/PhysRevD.101.103524, https://arxiv.org/abs/2002.02695

20. Cicoli M, Dibitetto G, Pedro FG (2020) Out of the swampland with multifield quintessence? JHEP 10:035. https://doi.org/10.1007/JHEP10(2020)035, https://arxiv.org/abs/2007.11011
21. Boyle LA, Caldwell RR, Kamionkowski M (2002) Spintessence! New models for dark matter and dark energy. Phys Lett B 545:17. https://doi.org/10.1016/S0370-2693(02)02590-X, https://arxiv.org/abs/astro-ph/0105318
22. Brinkmann M, Cicoli M, Dibitetto G, Pedro FG Stringy multifield quintessence and the swampland. https://arxiv.org/abs/2206.10649
23. Abdus Salam S, Abel S, Cicoli M, Quevedo F, Shukla P (2020) A systematic approach to Kähler moduli stabilisation. JHEP 08:047. https://doi.org/10.1007/JHEP08(2020)047, https://arxiv.org/abs/2005.11329
24. Cicoli M, Quevedo F, Savelli R, Schachner A, Valandro R (2021) Systematics of the α' expansion in F-theory. JHEP 08:099. https://doi.org/10.1007/JHEP08(2021)099, https://arxiv.org/abs/2106.04592
25. Cicoli M, Cunillera F, Padilla A, Pedro FG (2022) Quintessence and the Swampland: the numerically controlled regime of moduli space. Fortsch Phys 70. https://doi.org/10.1002/prop.202200008, https://arxiv.org/abs/2112.10783
26. Romans LJ (1986) Massive N=2a supergravity in ten-dimensions. Phys. Lett. B 169:374. https://doi.org/10.1016/0370-2693(86)90375-8
27. Dyson FJ (1952) Divergence of perturbation theory in quantum electrodynamics. Phys Rev 85:631. https://doi.org/10.1103/PhysRev.85.631
28. Aoyama T, Hayakawa M, Kinoshita T, Nio M (2012) Tenth-order QED contribution to the electron g-2 and an improved value of the fine structure constant. Phys Rev Lett 109:111807. https://doi.org/10.1103/PhysRevLett.109.111807, https://arxiv.org/abs/1205.5368
29. Hanneke D, Fogwell S, Gabrielse G (2008) New measurement of the electron magnetic moment and the fine structure constant. Phys Rev Lett 100:120801. https://doi.org/10.1103/PhysRevLett.100.120801, https://arxiv.org/abs/0801.1134
30. Bender CM, Wu TT (1969) Anharmonic oscillator. Phys Rev 184:1231. https://doi.org/10.1103/PhysRev.184.1231
31. Banks TI, Bender CM (1972) Anharmonic oscillator with polynomial self-interaction. J Math Phys 13:1320. https://doi.org/10.1063/1.1666140
32. Bender CM, Wu TT (1973) Anharmonic oscillator. 2: A study of perturbation theory in large order. Phys Rev D 7:1620. https://doi.org/10.1103/PhysRevD.7.1620
33. Graffi S, Grecchi V, Simon B (1970) Borel summability: application to the anharmonic oscillator. Phys Lett B 32:631. https://doi.org/10.1016/0370-2693(70)90564-2
34. Palti E, Vafa C, Weigand T (2020) Supersymmetric protection and the swampland. JHEP 06:168. https://doi.org/10.1007/JHEP06(2020)168, https://arxiv.org/abs/2003.10452
35. Akrami Y, Kallosh R, Linde A, Vardanyan V (2019) The landscape, the swampland and the era of precision cosmology. Fortsch Phys 67:1800075. https://doi.org/10.1002/prop.201800075, https://arxiv.org/abs/1808.09440
36. Hebecker A Lectures on naturalness, string landscape and multiverse. https://arxiv.org/abs/2008.10625
37. Kachru S, Kim M, Mcallister L, Zimet M (2021) de Sitter vacua from ten dimensions. JHEP 12:111. https://doi.org/10.1007/JHEP12(2021)111, https://arxiv.org/abs/1908.04788
38. Kachru S, Kallosh R, Linde AD, Trivedi SP, De Sitter vacua in string theory. Phys Rev D 68:046005. https://doi.org/10.1103/PhysRevD.68.046005, https://arxiv.org/abs/hep-th/0301240

Part II
String Models

Chapter 7
Corrections to the Supergravity Action in Type IIB

In this chapter we describe the most common and best understood ways to break the no-scale structure for the tree-level scalar potential described in Chap. 2. Moduli stabilisation and the effects of perturbative and non-perturbative corrections to the tree-level action are best understood in type IIB supergravity. Therefore, we will limit ourselves to the discussion of type IIB constructions. For a review on the pros and cons of type IIB models see [1].

We begin by reviewing perturbative and non-perturbative corrections to the tree-level type IIB effective field theory in Sect. 7.1. In the subsequent sections, we dedicate some time to motivating the origin of the corrections that we consider from the point of view of the ten-dimensional type IIB theory. In Sect. 7.7, we discuss the regime of validity of the four-dimensional description. Finally, we introduce de Sitter model building within the context of: the KKLT scenario in Sect. 7.8, the racetrack scenario in Sect. 7.9 and the large volume scenario in Sect. 7.10.

7.1 Review of Type IIB Effective Field Theory

We begin with a brief review of the main techniques for deriving the form of the underlying scalar potential of string compactifications, with a view to building a robust model of dynamical dark energy with all moduli suitably stabilised. More detailed reviews can be found in [2, 3]. We assume that the potential is given by the F-term expression

$$V = \mathrm{e}^K \left[K^{i\bar{j}} D_i W D_{\bar{j}} \bar{W} - 3|W|^2 \right], \tag{7.1}$$

where $D_i W = (\partial_i + \partial_i K)W$ is the Kähler covariant derivative and $K^{i\bar{j}}$ is the inverse of the Kähler metric $K_{i\bar{j}} = \partial_i \partial_{\bar{j}} K$. Our focus will be on type IIB string compactifications in which the complex structure moduli and the dilaton are fixed at semiclassical level, and so ∂_i denotes partial differentiation with respect to the Kähler moduli

F. Cunillera, *Dark Energy*, Springer Theses,
https://doi.org/10.1007/978-3-031-21468-4_7

$T_i = \tau_i + i\theta_i$. Even though our focus here is on the effective action of type IIB string theory, our final phenomenological considerations on quintessence also apply more generally to type IIA and heterotic setups.

At tree-level, we have a Kähler potential, $K = K_0 - 2\ln\mathcal{V}$ and a superpotential $W = W_0$, where $\mathcal{V}$ is the volume of the internal Calabi-Yau. K_0 and W_0 include the complex structure moduli and the dilaton that have already been stabilised, and are therefore assumed to be constant. Because of the 'no-scale structure', the corresponding scalar potential vanishes identically. Therefore, to generate the appropriate masses for the Kähler moduli, we must include at least one of the following: (i) perturbative corrections to the Kähler potential, $K \to K + \delta K_{\rm p}$; (ii) non-perturbative corrections to the superpotential, $W \to W + \delta W_{\rm np}$; (iii) higher derivative corrections to the scalar potential, $V \to V + \delta V_{\rm hd}$.

A general formula giving the Kähler moduli dependence of perturbative and higher derivative corrections at all orders in α' and g_s has been provided in [4] exploiting a combination of higher dimensional symmetries such as supersymmetry, scale invariance and shift symmetry, together with techniques from F-theory. This formula reproduces several known explicit computations of quantum corrections. Here we focus on those which have been used for cosmological applications:

- α'^3 *corrections*
  These are perturbative corrections in $\ell_s = 2\pi\sqrt{\alpha'}$ to the Kähler potential. The leading one arises from $\mathcal{O}(\alpha'^3)\mathcal{R}^4$ terms in the ten-dimensional action and looks like [5]

$$K \to K_0 - 2\ln\left(\mathcal{V} + \frac{\hat{\xi}}{2}\right), \tag{7.2}$$

  with $\hat{\xi} = -\frac{\chi(\mathcal{X})\zeta(3)}{2g_s^{3/2}(2\pi)^3}$ where $\chi(\mathcal{X})$ is the Euler number of the Calabi-Yau $\mathcal{X}$.
- *Open string 1-loop corrections*
  These are corrections in g_s to the Kähler potential, $K \to K + \delta K_{g_s}$, and are conjectured to take the form [6, 7]

$$\delta K_{g_s} = \sum_i g_s \frac{C_i(U,\bar{U})t_i^{\perp}}{\mathcal{V}} + \sum_i \frac{\tilde{C}_i(U,\bar{U})}{t_i^{\cap}\mathcal{V}}. \tag{7.3}$$

  Here there are two contributions: those of $\mathcal{O}(g_s^2\alpha'^2)$ coming from the tree-level exchange of Kaluza-Klein closed strings, with $t_i^{\perp}$ denoting the 2-cycles perpendicular to the branes; and those of $\mathcal{O}(g_s^2\alpha'^4)$ coming from winding strings, with $t_i^{\cap}$ denoting the 2-cycles of the intersection among branes. C and $\tilde{C}$ are unknown functions of the complex structure moduli U, although, as the complex structure sector is fixed at tree-level, one can consider them to be constants.
- *Higher derivative corrections*
  These are also α'^3 corrections to the scalar potential arising from the dimensional reduction of ten-dimensional higher derivative terms of the form $\mathcal{R}^2 G_3^4$, that yield $V \to V + \delta V_{\rm hd}$ with [8]

U

V